ESCAPING
THE RABBIT HOLE

How to Debunk Conspiracy Theories
Using Facts, Logic, and Respect

MICK WEST

Skyhorse Publishing

Skyhorse Publishing books may be purchased in bulk at special discounts for sales promotion, corporate gifts, fund-raising, or educational purposes. Special editions can also be created to specifications. For details, contact the Special Sales Department, Skyhorse Publishing, 307 West 36th Street, 11th Floor, New York, NY 10018 or info@skyhorsepublishing.com.

Skyhorse® and Skyhorse Publishing® are registered trademarks of Skyhorse Publishing, Inc.®, a Delaware corporation.

Visit our website at www.skyhorsepublishing.com.

10 9 8 7 6 5 4

Library of Congress Cataloging-in-Publication Data has been applied for.

Cover design by Brian Peterson

Print ISBN: 978-1-5107-5577-2
Ebook ISBN: 978-1-5107-3581-1

Printed in the United States of America

Contents

Acknowledgments

The backbone of this book is the collection of accounts from people who have escaped from the rabbit hole. I am very grateful to each of them for the time they took to share their stories with me, and for their continued resilience. My thanks go to Willie, Martin Beard, Steve, Stephanie Wittschier, Karl, Richard, and Bob. Thanks also to the many other people whom I interviewed or simply chatted with online. While I did not get everyone's story in, their input greatly informed this book. Many of them I only know by internet pseudonyms, or they wished to remain anonymous, so I shall list only first names, thanking Nathan, Elliot, Scooter, Michael, Adam, Julie, David, Ed, Marty, Frank, Johnny, DJ, and Joshua.

Thanks to the members of Metabunk, my main forum for discussing these topics. I am particularly grateful to Deirdre, both for useful feedback on my communication style, and invaluable help in moderating Metabunk while I was distracted with writing. Thanks also to Trailblazer, Pete Tar, Landru, Trailspotter, TEEJ, JFDee, Efftup, Jay Reynolds, TWCobra, MikeC, NoParty, Ross Marsden, Scombid, Skephu, Whitebeard, MikeG, and many others.

Thanks to the prolific skeptical author Benjamin Radford, who took the time to read an early draft and gave many valuable suggestions. I just wish I'd asked him earlier!

Thanks to Joe Rogan, a rabbit hole escapee himself, who invited me on his show to explain Chemtrails, then later Flat Earth and other conspiracies, helping create the platform for this book.

Thanks to my agent Jill Marr and my editor Andrew Geller for their support and guidance through the entire process. Special thanks must go to Andrew for taming my random mid-Atlantic spelling and punctuation into some kind of consistency.

My deepest thanks must go to my wife, Holly. For many years she has been supportive of my vague ramblings about someday writing a book. The fact that I have finally produced something is in large part due to her encouragement, her constructive suggestions, and her motivating beratements. As an accomplished author herself, she helped me navigate the obstacles and pitfalls of both writing and publication. Thank you, Holly, for being my partner both in work and in life.

Willie – Rabbit Hole Escapee

When Willie first discovered the conspiracy theory rabbit hole he was a young man living in the Pacific Northwest. His hometown was, as he puts it, "a very liberal, hippy-dippy kind of community." He listened to Art Bell, an apocalyptic conspiracy theorist on short-wave radio, and he'd "wake up every morning thinking the world was going to end."

Willie got his information about what was going on in the outside world from a variety of sources. The most significant was What Really Happened, a website with the tagline, "The history the government hopes you *don't* learn!" Here's Willie:

> So I started every morning, reading "the news" on What Really Happened. It was all just that kind of conspiracy thing, and every once in a while they would link to one of those other sites, like Alex Jones' Infowars . . . And, as with a lot of people, I was just like, "Oh my God! Look at all this news that no one else has ever seen before, and I'm seeing it!"

Willie was an avid consumer of this special type of news, reading about all kinds of conspiracy theories. He read the theory that TWA Flight 800 was shot down by a missile, and conspiracy theories about the Waco siege and the Oklahoma City bombing.[1] He read how the government was planning to shepherd people into concentration camps, and how they added fluoride to the water to keep us weak. He read about the JFK assassination being a CIA plot and how the attacks on 9/11 were an "inside job." He read and believed many theories in the years he spent down the rabbit hole.

He'd heard about the Chemtrail theory early on, but he didn't pay much attention. Generally speaking, the theory says that the trails that planes leave behind in the sky are not, as science tells us, just condensation clouds, but are actually some kind of deliberate toxic spraying. These theories date back to 1997 when the focus was on the health effects of the supposed chemicals. These concerns didn't resonate much with Willie, a healthy young man, so he didn't really look into it.

What eventually drew him deep down the Chemtrails corner of the rabbit hole was a photo of "Chemtrail tanks" on a plane. A popular example of evidence used by promoters of the Chemtrail theory is photos of suspicious looking metal barrels on planes with tubing coming out of them. These are actually just photos of the interiors of test aircraft. The barrels contain water, used as ballast to simulate the weight of the passengers for flight testing. But if you don't know this then these photos could easily look like evidence of some kind of spraying campaign.

> The Chemtrail theory was in the background for quite a long time; I believed that 'they,' the government, or somebody, was capable of something like that. But I never said, "I'm feeling health effects from Chemtrails" or anything like that. The smoking gun for me, [in support of] Chemtrails, was the barrels thing. When I saw that I was like, "Oh well, that proves it, oh my God." I was somewhat devastated because it confirmed that it was true.

Willie stayed down the rabbit hole for years, consuming all the information he could find about conspiracy theories, and occasionally sharing the information in the comments section of his local newspaper. We often think of conspiracy theorists as ardent campaigners for what they think is the truth—evangelical proselytizers who harangue everyone with their alternative ideas. But many of them largely keep their theories to themselves.

> I didn't really go around proselytizing, but I did sign my comments on the local newspaper website with a little blurb about 9/11 being an inside job. But it wasn't like I was standing with a microphone. I was in a band, and I never said this onstage or anything like that.

When I asked Willie how he dealt with people around him who tried to dissuade him from his conspiracy beliefs, it turns out the situation was remarkably infrequent. He was living in something of a cultural bubble.

No, no one argued! Where I live there's a lot of people who believe in this stuff; I guess that says a lot about how I have a limited group of friends, or whatever. But I never really got any pushback. It was like I got radicalized online.

Maybe one time someone challenged me, and I gave them the [sarcastic] line, "Sure, governments always tell the truth." He came back to me with, "Well, no that's not true. Yes, governments lie, but in this case, there isn't really convincing evidence of what you say." The people on the local newspaper website comments section who were always commenting to me, I thought of them as way too rational. Not enough "feeling," you know? They were just too data-driven, mostly Apollonian in thinking rather than Dionysian, you know?

Apollonian and Dionysian refers to two Greek gods. Apollo is the god of rational thinking, prudence, and purity. Dionysus is his opposite—irrational, chaotic, a risk taker, driven by emotions and instincts. He's the fun guy. The devil-may-care Dionysian worldview was more appealing to the hippies of the Pacific Northwest.

Not only did Willie get very little critical pushback from the people around him, he wasn't even really aware that there *were* any places he could go to get information that countered the conspiracy theories he was hearing from his own "news" sources.

No, I wasn't aware of debunking websites in the beginning. In the beginning I was looking at my local news and eager to discuss world views with people, but I didn't know about Snopes, or . . . what others are there?

But, years later, it was the very same evidence that pulled him down that helped him escape: the ballast barrels.

One day, I was on Above Top Secret [a conspiracy forum], and somebody debunked the ballast barrel thing. When they gave me the information, a link to the aviation site where the image was from, I looked at it and thought, "Oh my God, the other thinking was the wrong thinking." I immediately recognized that the person who was doing this debunking on ATS was way smarter than me, more intelligent than me, way closer to being an expert on this subject than me. It was like an "aha, eureka!" moment, like "wait a minute." Somebody has been lying about these ballast barrels, to make it seem like the whole Chemtrail thing is real.

This "aha" moment, triggered by a friendly poster online, led Willie to my debunking forum, Metabunk, where I have a long thread that discusses and explains all the photos of ballast barrels that get passed off as "Chemtrail" barrels.[2] After this experience, Willie gradually shifted away from his unquestioning belief in alternative accounts of "the news," and began along a more skeptical, Apollonian path of questioning both sides of what he was being told. After he figured out that Chemtrails were not real he started to question other theories he had assumed were true, like explosives being used to destroy the World Trade Center. He'd made a U-turn deep down in the rabbit hole and was finally returning into the light. It had taken nine years.

About 2003 is maybe when I started believing all the weird stuff, until about 2012 or so. I was into all that stuff for a long time. Now I check out Metabunk all the time, just to see what the new thing is. I've been posting stuff on Facebook for my friend. I've adopted this skeptical frame of mind when looking at anything. I really . . . it's definitely changed my life.

I don't really think any conspiracy theories are really plausible. Even with "who killed Kennedy?"—there's these new documents just released, and there's no smoking gun, so I just don't give it much time or energy. I'm pretty satisfied with the official story. There's some weird stuff, but I'm pretty satisfied with it. Even the global warming hoax theory doesn't do it for me. I put it this way: I used to be entertained by conspiracy theories, but now I'm entertained by seeing them debunked.

In my circle people are confused by me now, they are like, "Are you a Republican now?" Because I don't believe 9/11 was an inside job now people can't quite figure me out.

"Are you a Republican now?" his friends asked him when he stopped thinking 9/11 was an inside job. Conspiracy theories often have a distinct partisan flavor, which we shall discuss later in the book.

> *And, you know, I really do appreciate the politeness aspect of Metabunk, there's so many debunking resources out there that belittle people and call them stupid and stuff.*
>
> *The conspiracy thing was a worldview. I was testing that worldview out. It was entertaining. It also feels good because you think you have the truth and nobody else does . . . But yeah, that's how it happened for me, and then one day, it might have been you on Above Top Secret, the ballast barrels, totally changed my life. Thank you.*

If it was me on Above Top Secret I was probably one of several people who posted the correct information. There were many skeptical types on ATS back then, debunking Chemtrails, unaware at that moment that they had changed someone's life just by showing them some information they had been missing.

Willie's story demonstrates that people deep down the rabbit hole can still escape. But it also shows us just what a different world it is down there. Willie did not simply reject the conventional explanation of things. He was, in many cases, simply unaware that a conventional explanation existed, except as some kind of abstract, too-rational lie that he instinctively stayed away from. For Willie the first step was not being persuaded that his position was wrong, it was being shown that other positions even existed and that good, intelligent people actually took them seriously.

Willie stayed down the conspiracy theory rabbit hole for so long because he was surrounded by like-minded people. Exposure to missing information gradually altered his perspective on the world and helped him escape. He found this missing information piecemeal via online interactions, but people get out much quicker if they are helped by a friend, someone who could introduce them to new perspectives in ways a stranger on the internet never could.

Introduction

Conspiracies are very real, of course. The fact that powerful people make secret plans at the expense of the general public should come as no surprise to anyone. Nixon conspired to cover up Watergate. The CIA staged "false flag" operations in 1953 to bring down the Iranian government. Powerful men in the Reagan administration conspired to illegally trade arms with Iran to finance the Nicaraguan Contras. Enron conspired to shut down power stations to raise the price of electricity. Executives from Archer Daniels Midland conspired to fix the price of animal feed.[1] People within the second Bush administration conspired to present sketchy evidence as conclusive proof of WMDs to justify the invasion of Iraq. Politicians tacitly (and sometimes overtly) conspire with wealthy individuals and corporations, helping pass favorable legislation in exchange for campaign contributions, or sometimes just bribes.[2] The prison industry conspires to get those politicians to incarcerate more people simply to maximize their profits.[3]

Nobody sensible is denying that conspiracies happen. These are well documented and undisputed facts. Conspiracies very clearly have happened and will continue to happen. Nobody is asking you to trust that the people in power always have your best interests at heart, because they clearly do not. Nobody is asking you to blindly trust the government, or big pharma, or any large entity with a gross amount of power, wealth, and influence. A key aspect of a well-functioning democracy is that the government should be subject to scrutiny.

Conspiracies are real, but with every one of these very real conspiracies and plausible potential conspiracy there's a slew of *false conspiracy theories*.

These theories are efforts to explain some event or situation by invoking a conspiracy. They are theories that are either very likely false because they lack the significant evidence needed to improve over the conventional explanation, or are simply demonstrably false.

There are conspiracy theories like the idea that the World Trade Center towers were destroyed with pre-planted explosives, or that the Moon landings in the 1960s were faked, or that planes are spraying toxic chemicals to deliberately modify the climate. There's less extreme, but still false conspiracy theories, like the pharmaceutical industry covering up how well homeopathy works (it doesn't), or the car industry covering up motors that can run on water (they can't). At the far end of the conspiracy spectrum there's the claim that the Earth is flat (it's not) and the government is somehow covering this up (how would that even work?). There's old conspiracy theories, like the idea that Jewish bankers rule the world, and new conspiracy theories, like the idea that the government stages shootings of children in schools to promote gun control.

The premise of this book is very simple. These false conspiracy theories are a problem. They hurt individuals by affecting their life choices, in terms of money, health, and social interactions. They hurt society by distracting from the very real problems of corruption and decreasing citizens' genuine participation in democracy. False conspiracy theories are real problems and we can and should do something about them. This book discusses the nature of the problem, why people get sucked in, how they get out, and what pragmatic things can be done to help individuals escape the conspiracy theory rabbit hole.

The key themes of this book are:
- Understanding the conspiracy theory rabbit hole
- Realizing that conspiracy theorists are just normal people
- Developing a clear understanding of what they are thinking and why
- Fostering trust and mutual respect
- Finding areas of agreement and recognizing their genuine concerns
- Identifying mistakes in their beliefs, or areas where they lack information
- Exposing them to new information to help them gain a more fact-based perspective

- Doing it all with honesty and openness
- Giving it time

In this book I will draw on three sources of information. Firstly and primarily, I will draw upon my personal experience. I run the website Metabunk, which is a site for discussing, investigating, and debunking a wide variety of false conspiracy theories and unusual beliefs. With my previous "Chemtrail"-focused site, Contrail Science, and other sites, I've been debunking as a hobby for over fifteen years. During that time, I've met hundreds of people on both sides of the fence, heard their stories, and seen them change over the years. Many of them I've helped, usually indirectly, like Willie with the Chemtrail barrels, but sometimes directly. I'll include the stories of more of those people in later chapters.

Secondly, I will draw upon the writings of other skeptical-minded people doing the same thing. In fields ranging from global warming conspiracy theories to 9/11 conspiracies, there are others who do similar things to myself, people who have researched both why people believe conspiracy theories and how they can be helped out. Many individuals have shared their experiences and thoughts about which debunking and communication strategies work, and which do not.

Thirdly, I will draw upon the academic literature in the field of conspiracy theories. Since the 1950s with the conspiracy theories of the radical right, through the 1960s with the assassinations of JFK, RFK, and MLK, and especially after 9/11 in 2001, there has been a steady growth in interest in why people fall for unfounded beliefs and what strategies are scientifically effective in bringing them back to a more realistic view of the world.

The fundamental technique outlined here is maintaining effective communication and presenting your friend, the conspiracy theorist, with information that they are lacking, and doing it all in a manner that will encourage them to look at what you are presenting without rejecting you as an idiot or a government shill. Given time, this additional information will help them gain enough genuine perspective to begin to question what they thought they knew and to start their journey out of the rabbit hole.

What Is the "Rabbit Hole"?

The normal definition of the metaphorical rabbit hole is something like:

> *A bizarre world, a time-consuming tangent or detour, often from which it is difficult to extricate oneself.*[4]

In the modern conspiracy culture this rabbit hole is an obsession with a bizarre world of books, websites, and YouTube videos that claim to reveal hidden truths about the world. It's a detour from regular life, one that is certainly time-consuming, and definitely one from which it is difficult to extricate oneself.

The phrase comes from Lewis Carroll's *Alice's Adventures in Wonderland*. Alice enters the bizarre Wonderland by following a white rabbit down a hole.

> *Down went Alice after [the White Rabbit], never once considering how she would get out again. The rabbit-hole went straight for some way, and then dipped suddenly down, so suddenly that Alice had not a moment to think about stopping herself before she found herself falling down what seemed to be a very steep well.*[5]

In recent times a more specific usage has arisen, derived from the 1999 film *The Matrix,* where at a crucial point Morpheus (Laurence Fishburne) offers Neo (Keanu Reeves) a choice. He can either take the blue pill and return to a normal life, or take the red pill and "see how deep the rabbit hole goes."

Neo, of course, "takes the red pill," and the "rabbit hole" leads him to discover the true nature of the world. He "wakes up" from his programmed illusion of comfortable, bland monotony into a brutal yet genuine struggle for existence, a messianic battle against evil, manipulating overlords.

This terminology has been directly adopted by various conspiracy communities. The rabbit hole is seen as a good place to be, a place where the true nature of the world is revealed. Their red pill moment might be the first YouTube video they watched, a conversation with a friend, or a book. They wake up, take the red pill, and proceed deliberately down the rabbit hole into what they see as an incredible wonderland of truth.

I also want people to wake up to the true nature of the world. But the conspiracy theory rabbit hole is not the way to do it. It's full of seductive nonsensical theories, a bizarre wonderland of time-wasting and harmful falsehoods that are taking people further away from the real world, not closer. It's not a blue pill or a red pill; it's a poison pill.

This book is about helping people out of that rabbit hole of false conspiracy theories. More specifically it's about helping your friend.

Your Friend

This book is written mostly assuming that you, the reader, are trying to better understand or help someone who is down the rabbit hole. Perhaps it's a relative, maybe your spouse, a child, a parent, a sibling. Perhaps it's a friend, a close friend or casual acquaintance, someone you sit next to at work, or even just someone you know online. With this in mind I'm going to refer throughout the book to this person—the target of your concerns and your attention—as "your friend."

Of course, they might not currently be your friend. Especially in family situations, a strong belief in something that another person finds preposterous can lead to frustration, anger, and possibly even to deep-seated animosity or disgust. Your friend might find it ridiculous that you think people landed on the Moon. He might consider you borderline insane for entertaining the notion that Lee Harvey Oswald acted alone. He might grow angry when you refuse to watch all four hours of *September 11: The New Pearl Harbor*. He might turn his back on you when you refuse to be concerned about the white lines crisscrossing the skies.

But if you wanted a book for dealing with an enemy, a list of tricks you can use to annihilate someone in a debate, something that will make them look like an idiot, then I suggest you look elsewhere. I want to *help* people, not mock or belittle them. If you think you can only help someone by beating them in every argument and making them look stupid, then I respectfully disagree. Showing your friend their faults is only a small part of helping them out of the rabbit hole, and if you apply such a blunt tool to someone you consider your enemy, then you will probably achieve the opposite of your goal, only hammering them deeper and deeper into the rabbit hole as they harden their heart against you and their mind against your facts.

So even if they are actually acting in some sense as your enemy, I will still refer to them as "your friend." Try to think of them as such, a good person who means well, someone who is simply mistaken about certain things and rather set in their ways. As we will shortly see, the first stage of helping someone out of the rabbit hole is to understand them, and then to gain their trust. You cannot do that by waging a war of words against them.

There is a significant danger that I will reiterate throughout the book. The danger is that advice like "treat them like a friend" and "gain their trust" might be viewed as advice from a manual on brainwashing. Conspiracists are obviously suspicious of people like me who spend time investigating and refuting their theories. I get accused of being a paid government agent, someone trained in "disinformation," someone skilled in implanting false ideas in people's heads. They may look at this book, and my body of work on Metabunk, and decide I'm lying, trying to gaslight individuals away from the truth.

The best defense is to be as open and honest as possible. Yes, I think treating someone like a friend makes it easier to convince them of their errors. But the only reason they are acting like an enemy towards me is because they are mistaken in their beliefs. If I'm engaging with someone it is because I think they are a good person who is just stuck down a rabbit hole. If they think I'm the enemy, and they act as my enemy, then it's only because they are in fact a friend who has lost their way.

Finally, "your friend" might in fact be *you*. Perhaps you are reading this book because you recognize you are a little lost down a rabbit hole and you want a little help out, or at least a look outside. Perhaps you don't think you are down the rabbit hole, or you think that your beliefs illustrate you are wide awake to the truth. Perhaps you are reading this because you think I'm a government shill, and you want to get the lowdown on this new government shill handbook so you can help your friends avoid being tricked. Or maybe someone asked you to read this book as a favor, and you begrudgingly agreed, because they are your friend.

If you are actually a conspiracy theorist, then you can think of "your friend" in one of two ways. Firstly, you should be your own friend. You might start out reading this book to try to figure out my mind games, but I hope you end up with some better perspective on both where I am coming from, and on your own view of how the world works. Maybe you'll find you've got something wrong somewhere. Maybe you will at least find this perspective helps you better communicate your own ideas. Maybe this book will confirm what you already knew. Whatever the outcome, I hope you find it useful.

The second way a conspiracy theorist might find this book useful comes about because conspiracy theories exist on a spectrum. If you are a conspiracy theorist (and we all are to some degree), you consider yourself a reasonable person, and you believe only in conspiracy theories that you think are well founded, backed up by evidence and common sense. While you might disagree with my attempt to debunk your theories at wherever level you are at on the conspiracy theory spectrum, you might find common ground in trying to help those who are further along. I've had several 9/11 Truthers thank me for helping debunk Chemtrails, and I've had Chemtrail believers thank me for explaining to their friend why the Earth is not flat. Read this book to figure out how to help your friend who's down a deeper, darker rabbit hole. If it seems reasonable then maybe at some point you can see if anything in here applies to your personal beliefs.

Or, if you like, go ahead read this as a brainwashing manual for government shills. Try to figure out my tricks. I'm not trying to brainwash you, but if it will get you to read the book then go ahead and assume it for a while, but I invite you to check back again later.

What's the Harm?

"Why bother?" is a question I am asked a lot. Why should we care about people who believe in conspiracies, and why should we try to help them? This question speaks directly to the reasons why I wrote this book. I want to help people out of the conspiracy theory rabbit hole because *false conspiracy theories cause harm*. They do so in several ways.

Perhaps most significantly, there is harm at a direct individual level, the level of your friend. If they believe that the efficacies of natural remedies (homeopathy) are being covered up by large pharmaceutical companies then they might be tempted to avoid conventional treatments, and instead opt for herbal remedies that are not proven to work. In some cases this can lead to death.[6] If they think planes are spraying poison in the sky then they might waste their money on Chemtrail detox pills.[7]

There is also harm for the individual in their relationships, romantic or otherwise. A common result of belief in false conspiracy theories is marginalization and social isolation.[8] The rabbit hole becomes an obsession, and if one partner does not share that same obsession then significant relationship problems can develop, including divorce.[9] These problems extend to family and friends, and even into the workplace.[10]

Beyond the harm that a belief in false conspiracy theories brings to the individual and those around them, it can also lead to harm to others. Scientists researching the climate have been harassed and threatened by people who believe that they are covering up a conspiracy, even to the extent of receiving death threats.[11] Politicians have been heckled by 9/11 "inside-job" conspiracy theorists.[12] The parents of murdered children have been stalked by people who think they are part of a hoax.[13] One man fired a gun in a pizza parlor where he thought children were being held captive as part of a pedophilia ring involving the Clintons.[14]

Even more significantly, conspiracy theories can lead to major acts of terrorism, both domestic and foreign. Timothy McVeigh, the Oklahoma City Bomber, was a conspiracy theorist who thought a cabal of international Jewish bankers was taking over America.[15] Tamerlan Tsarnaev, the Boston Marathon Bomber, was part of a wave of radicalization of young Muslims, driven in large part by conspiracy theories spread via online videos.[16]

The practical harm is very real, very tangible. But there are less tangible aspects of the harm conspiracy theories leave in their wake. Truth matters in a society. The more that public discourse is based on falsehoods, the harder it is to make constructive progress. Science is harmed when there are millions of people who think that scientists are corrupt corporate shills. The democratic process is harmed when people vote based on their belief in conspiracy theories. The nation suffers when policies are enacted based partly on false claims. The international standing of our country is harmed when conspiracy theories are increasingly accepted by the general population.

So I bother, I debunk, to stave off and spare others from this harm. I encourage other people to do so, to help their own friends and, while it is perhaps just a drop in the ocean, to make the world a better place.

Can People Change?

Is it even possible? When I tell people that I debunk false conspiracy theories their reaction is sometimes, "But they never change their minds." Indeed, conspiracy theorists do often seem remarkably entrenched in their beliefs, able to withstand countless hours of reasonable rebuttals without giving an inch.

This is even an assessment conspiracists make about themselves. I joined a Facebook group called "9/11 Truth Movement" and announced I was looking for "Former 9/11 Truthers." I got a few responses from actual former Truthers, but I also got a lot of replies like this one:

> I'll tell you right now, I guarantee you will not find one "former Truther." Guaranteed, or the person is lying. Period. There's no unlearning that a crime like that was committed and covered up. To be convinced otherwise would require a well below room temperature IQ, which means you probably never doubted the official story to begin with.

First they say I'd never find any. Then they say if I did, all that means is someone was lying. Then they say that you'd have to be really stupid to stop being a Truther in the first place and since stupid people don't become Truthers that would be impossible. This attitude was surprisingly common among the group. For a true believer, no true believer would ever change their beliefs; it was literally impossible. If you pointed out people who had changed their minds and spoken publicly about it, they denounce them as shills or "gatekeepers," or say they never really believed in the first place.

But people *can* change, and I have found them (or they've found me). Over the course of several years, I've interacted with, spoken with, and met in person many current believers and many former believers. Some whom I helped get out of the rabbit hole. Conspiracy theorists may not think that change for themselves is ever possible, but it is, even for those who are the most convinced. The best way of illustrating that someone deep down the rabbit hole can get out is to consider stories of those who have done just that, people who were nearly as deep down the rabbit hole as one could go, and yet they got out. I opened the book with Willie's story as an immediate demonstration that change is possible. We shall meet other escapees in later chapters.

Why "Debunk"?

The word "debunk" is used throughout this book to denote the process of helping people understand why their conspiracy theories are not backed by good evidence. But the word "debunk" is sometimes interpreted to mean that the debunker has pre-judged the issue and is only interested in convincing others by whatever means necessary. So why use a term that might be perceived negatively?

I have discussed this with professional skeptic James Randi and veteran paranormal investigator Joe Nickell. Randi told me he thinks the term "debunker" carries too much of an assumption that the argument was presumed to be false, and as such he prefers to describe himself as a scientific investigator. Similarly, Nickell made the very compelling argument that what he does with each new case is not to set out to prove that it's not ghosts, but instead to investigate the circumstances, to see what actually happened, and then to explain it.

But I use the term "debunk" for two reasons. Firstly, I see debunking as a two-stage process of both investigating and then explaining. Debunking is defined as "exposing the falseness" of an idea or belief.[17] To expose a falseness you first have to find it, and then explain why it is false. When faced with a claim like "no plane hit the Pentagon on 9/11," you first look at the proposed evidence behind the claim, and investigate for factual accuracy. If you find inaccuracies then you can explain these to people.

Secondly, most people, including conspiracy theorists, have no problem with the use of the term in the past tense, such as "Claim X has been debunked." This is widely understood to mean that the claim has been investigated and conclusively shown to be incorrect.

But the focus of this book is not on investigating, it is on explaining. The majority of the conspiracy theory claims you will come across are not new claims that need investigating. They are old claims that have been investigated, and only persist in the minds of people like your friend because they are unaware of the most reasonable explanation or because they lack information that allows them to fully understand that explanation. Bringing those explanations and that missing information to your friend is what I mean by debunking.

Overview

Escaping the Rabbit Hole is arranged into three parts. In Part One, we take a detailed look at the conspiracy theory rabbit hole. Why do conspiracy theories exist? Why do people get sucked into them and how can you help them out?

Chapter One, "The 'Conspiracy Theory' Conspiracy Theory," addresses the contentious usage of the term "conspiracy theory" by looking at its history. The usage predates the assassination of JFK in 1963, and while it does have some negative connotations, I use it because it is a term that is (mostly) well understood.

Chapter Two, "Conspiracy Spectrums," looks at the range of conspiracy theories from the plausible to the ridiculous. I introduce the concept of a conspiracy "demarcation line" which is drawn on each individual's personal conspiracy spectrum. On one side of their line are the "reasonable" theories, and on the other are "silly" theories and "disinformation." I discuss how understanding and identifying this line is key to helping your friend.

Chapter Three, "The Shill Card," addresses the common accusation that some people who attempt to debunk false conspiracy theories are shills. The best way to combat this accusation is to be as honest and open as possible about what you are doing. To that end, this chapter contains a detailed explanation of how I ended up debunking conspiracy theories on the internet, why I do it, and how I can afford it.

Chapter Four, "The Rabbit Hole: How and Why," examines how people get sucked into the rabbit hole. What is the contribution of psychological factors? How do people typically end up in a conspiracy spiral? I look at current research on the matter and the significant role of online videos.

Chapter Five, "Core Debunking Techniques," lays out a set of tools and guidelines for practical debunking. The focus is on effective communication of missing information.

Chapter Six, "Steve – A Journey through the Rabbit Hole," tells the story of Steve, a conspiracy theorist since the 1970s whose story of escape exemplifies many of the concepts covered in the preceding chapters.

Part Two is the practical core of the book. Four different conspiracy theories are discussed in depth. The common false claims of evidence for those theories are described, and I explain how best to convey the explanations of those claims to your friend. Paired with each chapter is the account of someone who went down that particular rabbit hole and ultimately escaped.

Chapter Seven, "Chemtrails," covers the surprisingly popular idea that the white lines that planes leave in the sky are not just condensation, but are part of a secret plot to alter the climate. The science of contrails (what these white lines actually are) is covered, along with the most common claims like "contrails don't persist," "aluminum is in the rain," and "geoengineering patents." This is my personal favorite conspiracy theory, and the longest chapter in the book.

Chapter Eight, "Stephanie – A Former Chemtrailer," tells the story of Stephanie, a German Chemtrail believer who was helped out of the rabbit hole by her friend.

Chapter Nine, "9/11 Controlled Demolitions," looks at the most popular aspect of the wide range of 9/11 conspiracy theories—the idea that the plane impacts and raging fires were not enough to bring down the World Trade Center, and that pre-planted explosives must have been used. It's too large a subject to do justice in one chapter, so I focus on some key areas in which your friend might be missing information. These include: Architects and Engineers for 9/11 Truth, nanothermite, the plane that hit the Pentagon, and a (not) missing $2.3 trillion.

Chapter Ten, "Karl–Temporary Truther," tells the story of someone who nearly fell deep down the 9/11 rabbit hole, but was caught in time by his friend.

Chapter Eleven, "False Flags," covers the often emotive issue of claims that events like Sandy Hook were hoaxes. I take a detailed look at the often-cited historical evidence for False Flags, in particular Operation Northwoods and the Gulf of Tonkin incident. I look at some ways you can bring perspective to people who have fallen down this particular rabbit hole.

Chapter Twelve, "Richard – Drawing the Line at Sandy Hook," tells the story of a young man for whom the Sandy Hook hoax theory was the thing that ultimately helped turn him away from his conspiratorial thinking.

Chapter Thirteen, "Flat Earth," covers what many consider to be the obviously ridiculous theory that the Earth is flat and the government is covering this up. Many people who claim to hold this view are just trolling, but what do you do if you meet people who actually believe it? I look at the history of the theory, the most common claims, and some very straightforward ways of showing people that the Earth is not flat.

Chapter Fourteen, "Bob – Escape from Flat Earth," tells the story of someone who not only believed the Earth was flat, but had family members who believed it too.

Part Three takes a look at some of the additional complications that you may encounter while helping your friend and finishes with a look at the future of debunking.

Chapter Fifteen, "Complications in Debunking," first examines the common problem of explaining a complex subject to someone who is simply (through no fault of their own) incapable of quickly understanding it. Further complications arise when your friend is a close family member, with a different dynamic from a friend and potentially more significant long-term ramifications. I look at the issues raised by the Morgellons theory and offer some brief advice on dealing with mental illness. I conclude with considerations and guidance on how to navigate political disagreements that cross over into the conspiracy realm.

Chapter Sixteen, "The Future of Bunk and Debunking," is partly speculative, but is based firmly against the backdrop of the influence of disinformation in the 2016 election, and the repercussions that continue to this day. I look at how trolls and bots help spread conspiracy theories, and how it is probably going to get worse before it gets better.

I conclude on a hopeful note with a look at the tools being developed to fight online misinformation, and how this might help turn back the tide of conspiracism.

PART ONE

CHAPTER ONE

The "Conspiracy Theory" Conspiracy Theory

"Conspiracy theory" is a term that I've used extensively for a long time, and yet I initially struggled with it, and constantly tried to find alternatives.

The problem is that "conspiracy theory" (and "conspiracy theorist") is considered by many to be deliberately derogatory. The fact that "conspiracy theory" is on the cover of this book might lead some people to dismiss the book as an attempt to mock or belittle the people who believe such things. But if you look at a typical dictionary definition it will be something like:

A theory that explains a situation or event as resulting from a secret plot by some powerful group.

"Conspiracy theorist" is being simply defined as a person who believes a conspiracy theory. This is a perfectly reasonable definition that fits what 9/11 Truthers believe, or what JFK conspiracists believe, and what Chemtrailers, Moon landing hoaxers, Sandy Hook False Flaggers, and alien base cover-uppers all believe. They think that there was a secret plot behind something, and/or that there's been a secret cover-up of something.

But being correct in the literal sense does not make a word immune to being offensive. It's the applicability to the more esoteric theories that is offensive to the more mainstream conspiracists. The average person who simply thinks that the CIA assassinated JFK sees himself as a reasonable person and does not want to be associated with the odd people who think the Queen is a shape-shifting lizard. Similarly, the 9/11 Truther does not want to be thought of as a "tinfoil hatter" who worries that the NSA is beaming messages into his brain with radio waves.

But beyond this simple association, there's a deeper reason why conspiracists shy away from the label. That reason is itself a conspiracy theory—the theory that the term "conspiracy theory" was invented in 1967 by the CIA to discredit conspiracy theorists.

This "conspiracy theory" conspiracy theory points to a 1967 CIA document that surfaced in 1976 after a FOIA request from the *New York Times*. The document, titled "Concerning Criticism of the Warren Report," is a fascinating snapshot of the time. The CIA is concerned, for a variety of reasons, that there's a rising tide of unfounded conspiracy theories that are damaging the reputation of the CIA and the government. They suggest ways of countering them, but they *don't* suggest using the term "conspiracy theory" as the way to do so.

But people who might have that label applied to them (like people who think the World Trade Center was destroyed with explosives) feel that the document is very much about labeling them as a "conspiracy theorist" in an attempt to ridicule and sideline them. One of the main promoters of this theory is Dr. Lance DeHaven-Smith, who used it as the central thesis of his book, *Conspiracy Theory in America*, writing:

> *Thus the conspiracy-theory label has become a powerful smear that, in the name of reason, civility, and democracy, preempts public discourse, reinforces rather than resolves disagreements, and undermines popular vigilance against abuses of power. Put in place in 1967 by the CIA, the term continues to be a destructive force in American politics.[1]*

DeHaven-Smith admits that the document itself does not actually explicitly encourage usage of the term, and to get around this he embarks on a series of interpretive mental gymnastics, attempting to determine the hidden meaning in the CIA document. He analyzes it sentence by sentence, and sometimes word by word, projecting his interpretation upon it.

> *CIA Dispatch 1035-960 appears to be a straightforward memo with clear language and reasonable motives, but it is actually a subtle document, conveying many of its messages by indirection and implication. To grasp the nuances in the*

text requires a very careful reading. Some sections of the dispatch clearly have a surface meaning for ordinary readers, and a deeper, less obvious meaning for readers who are listening for, as it were, a second frequency, a hidden meaning. Multiple levels of meaning occur in various forms of speech. . . .

CIA Dispatch 1035-960 is not a Platonic dialogue . . . but it is a document written by spies for other spies, and spies know that, as a written document, it could fall into the wrong hands, as, in fact, it did because of the Freedom of Information Act request. So we should assume that the dispatch may contain some veiled meanings.

While DeHaven-Smith claims that the "conspiracy theory" label was "put in place in 1967 by the CIA," in fact the term had been in use for decades before that. The first usage dates back to 1870 with a theory about a conspiracy to physically abuse the criminally insane in mental asylums.[2] The term took hold in the United States as a description for a particular theory about the secession of the South from the Union and appears in several books around 1895, nearly seventy years before the CIA document.[3] It continued to be used in the early twentieth century, such as in the paper "The 'Conspiracy Theory' of the Fourteenth Amendment" in 1930.[4]

A decade before the CIA memo, and years before JFK's assassination, the term was actually in use in the United States in much the same way as it is now—as a descriptor for largely unfounded theories that seek to explain events with a nefarious conspiracy. At that time one of the main sources of such theories was the "Radical Right"—extreme-right religious and nationalist organizations like the Ku Klux Klan and the John Birch Society. In 1960, William Baum wrote in "The Conspiracy Theory of Politics of the Radical Right in the United States":

. . . acceptance of the reality of an omnipotent and demonic conspiracy is the most significant and distinctive ideological characteristic of the contemporary American extreme or radical right.[5]

Baum's work was quite influential and was repeated in several papers and books. In 1962, the year before the assassination of President Kennedy, Walter Wilcox wrote "The Press of the Radical Right,"[6] which includes an attempt to quantify the various types of conspiracy theories. In it he gave several examples:

- NAACP is operated by a New York Jew through Negro Fronts
- Fluoridation [of drinking water] brings people under control as a narcotic, not good for teeth
- Unemployment is increasing in US because trade is in the hands of an international cult
- Organized Jewry tried to sabotage the gospel message in the film *Ben Hur*
- California intelligence tests give a choice of two evils, making one seem right

These theories do not seem too dissimilar to those seen today. The water fluoridation theory is still in existence, and is generally a foundational belief of people who hold to the more esoteric theories, like Chemtrails. Wilcox went on to propose what was probably the first conspiracy theorist spectrum, a zero through seven scale of "commitment to conspiracy" which was a measure of the extent to which a given article in a press publication of the radical right was devoted to a conspiracy theory:

Commitment to Conspiracy Scale
7 Preoccupied with conspiracy
5 Conspiracy conspicuous
3 Conspiracy present
1 Hints at conspiracy
0 No clear evidence of conspiracy

Wilcox also included a non-rationality scale, which contains descriptions you might still apply to many writings on the internet today:

Non-rational Scale
7 Paranoiac overtones, confused, few or no credible facts
5 Polemic, shrill, credible facts few and heavily stacked
3 Heavily one-sided, credible facts present
1 Mildly one-sided, credible facts lightly stacked
0 No clear evidence of non-rationality

Wilcox draws a connection between the degree of non-rationality in a conspiracy theory and how committed the person is to that theory.

> *For instance, it is logical to assume that non-rationality correlates to a marked degree with the theory of conspiracy* . . .

Clearly the CIA did not invent the term. Nor did they even suggest that the term be used as a way of belittling people. They used "conspiracy theory" and "conspiracy theorist" only once each in the entire document:

> *Innuendo of such seriousness affects not only the individual concerned, but also the whole reputation of the American government. Our organization itself is directly involved: among other facts, we contributed information to the investigation. Conspiracy theories have frequently thrown suspicion on our organization, for example by falsely alleging that Lee Harvey Oswald worked for us. The aim of this dispatch is to provide material countering and discrediting the claims of the conspiracy theorists, so as to inhibit the circulation of such claims in other countries.*

The term is used simply as a descriptor. The CIA would obviously have been familiar with the anti-government radical right, as they would be familiar with any anti-government organization. They would also have been familiar with academic writings about the radical right and the use of the term "conspiracy theory."

To convey this to your friend, the first step is to show them that the term existed prior to both the CIA document and the JFK assassination. Then if they need more detail show them the actual writings by Wilcox and others that used it the year before JFK's death in much the same way it is used today. They may still be unconvinced, and a more thorough debunking might require an examination of the full text of the CIA document.

An additional step is to look at what happened to the term "conspiracy theory" *after* the JFK assassination, and after the CIA used it in the often-cited document. To investigate this, I used the online Newspaper Archive database to extract the total numbers of uses of the term "conspiracy theory" in newspapers for each year from 1960 to 2011 (the last year that Newspaper Archives has a significant number of scanned papers). I adjusted the number of uses relative to the total number of words printed that year and plotted a graph.

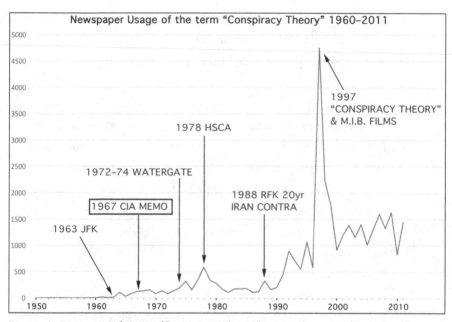

FIGURE 1: Frequency of usage of "conspiracy theory" in Newspaper Archives.

Clearly, if the CIA had intended to popularize the term after 1967, they failed. There were the few instances of the term before 1963 as already noted, but the first spike is actually in 1964 directly after the JFK assassination (November 22, 1963). The next year (1965) shows a dip, and then there's a steady increase in the subsequent years. In the year the CIA report was supposedly promoting the usage (1967) the term was *already* well established and was growing in popularity. You might have expected a surge in usage after the assassinations of Robert F. Kennedy and Martin Luther King Jr. in 1968; however, after 1969 it settled down.

There are spikes after that, a slow rise over the Watergate years of 1972 (when the Watergate break-ins happened) to 1974 (when President Nixon resigned). A big spike occurred in 1978 when the House Select Committee on Assassinations released its conclusions, including, "President John F. Kennedy was probably assassinated as a result of a conspiracy." Usage dropped back to previous levels in the 1980s with the exception of a minor jump in 1988, the twentieth anniversary of the RFK assassination and the year of the Iran–Contra scandal.

The 1990s are actually when the use of the term "conspiracy theory" really took off, increasing nearly 500 percent from 1990 to 1995 with the end of the Cold War, the start of the Gulf War, the LA Riots, the Waco siege, the start of *The X-Files*, and the Oklahoma City bombing.

There's a huge spike in 1997 with the releases of the films *Conspiracy Theory* and *Men in Black*. In both these films, as in most films about conspiracy theories, the theories turn out to be correct. There's obviously no Hollywood movement to belittle people with the term; in fact it's a very *positive* use. This is especially the case in the film *Conspiracy Theory* where the protagonist Jerry Fletcher (Mel Gibson) is at first seen as a crazy eccentric who is to be humored but ignored. As the film progresses it becomes clear that Jerry was actually correct, he was being watched by CIA agents, his theories were right, and eventually he becomes the hero.

After the film aired, the term "conspiracy theory" was firmly entrenched in American culture, and more generally in the English-speaking world. Subsequent developments simply built upon this. The "Chemtrails" theory was invented in 1998, and in 2001 we had the attacks on the World Trade Center and the Pentagon, immediately spawning a huge slew of theories.

Perhaps more significantly than the usages in popular culture, and perhaps even more significant than the events of 9/11, the late 1990s and early 2000s are when we saw the meteoric rise of the internet. Where Newspaper Archive leaves off in 2011, we can continue with other measures of the popularity of the phrase, such as Google Trends.

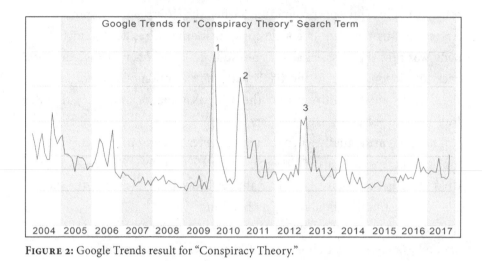

FIGURE 2: Google Trends result for "Conspiracy Theory."

This gives us a finer-grained view of interest in "conspiracy theories" and instead of being a measure of the mentions in newspapers, which is only an indirect measure of public interest, this gives us an actual measure of what the public was directly and actively searching information on. The term was declining in popularity until December of 2009 when the TV series *Conspiracy Theory with Jesse Ventura* was released (labeled "1"); this was followed by similar (but declining) peaks in October 2010 (Season 2) and November 2012 (Season 3). The show was hosted by Jesse Ventura (actor and former governor of Minnesota), and was again very positive in its portrayals of conspiracy theories, arguing strongly that most of the theories presented in the show were either true or at least reasonable things to be suspicious of. After the show ended, interest returned to pre-2009 levels, only picking up slightly around the 2016 election. The final spike shown in October 2017 was in response to the Las Vegas Massacre.

We see the history of the usage of the term is overwhelmingly dominated by *positive* associations in the popular media. The portrayals such as in *The X-Files* or Mel Gibson's *Conspiracy Theory* are honest in their recognition of the public perception of conspiracy theorists as eccentrics, and then almost always portray them as being the people who are correct. *The conspiracy theorist comes across as the hero*, someone who has accurately deduced some aspect of the inner workings of the world and is seeking to expose that secret. Instead of there being some deliberate program in the media to denigrate conspiracy theorists, the biggest usages of the term in the last twenty years are all in ways that might even be thought of as to be trying to rehabilitate and promote it.

While "conspiracy theory" does have some negative connotations, it has also given the conspiracy culture a degree of legitimacy that might otherwise be lacking. Consider that before the wider adoption of the term, one of the most influential essays on the topic was Hofstadter's 1964 piece "The Paranoid Style in American Politics," which used the far more directly insulting term "paranoids" to refer to those who tended to explain all events as the result of some conspiracy.

> *I believe there is a style of mind that is far from new and that is not necessarily right-wing. I call it the paranoid style simply because no other word adequately evokes the sense of heated exaggeration, suspiciousness, and conspiratorial fantasy that I have in mind.*[7]

If we didn't have the "conspiracy theorist" term, it's quite possible that the people we now call conspiracy theorists might equally have been called "paranoids" or some other directly pejorative term. By contrast the current label is relatively neutral.

What we have here is an asymmetry in perception. The conspiracists reject the accurate labels given to them because they think it's an attempt to belittle them. They do not consider their constant suspicions to be in any way unusual (except in contrast to the sheep-like acquiescence of the general public). But because their suspicions are generally unfounded and out of the mainstream, then any label their group acquires is going to eventually become perceived as derogatory.

DeHaven-Smith is an example of this asymmetry; he rejects the notion that the negative connotation of "conspiracy theory" might have anything to do with the generally baseless and often unfounded claims of most conspiracy theories, and argues that instead of "conspiracy theorist" one should use "conspiracy realist," and instead of "conspiracy theory" one should say "state crime against democracy" (SCAD).

He misses the point. If a group manages to get a label to stick then it's not going to change the public perception. Conspiracy theorists are not judged to be on the fringe because they are part of a group called "conspiracy theorists." They are on the fringe because they make unfounded, unrealistic, or overly speculative claims. Labels do not define the perception of a group; the labels *take on* that perception. After the UK Spastics Society was renamed "Scope" in 1994, the playground insult of "spastic" for a clumsy kid was simply supplanted by the insult "scoper."[8] If DeHaven-Smith could miraculously get large numbers of people to adopt "SCAD," then all that would happen would be that conspiracy theorists would also be called "Scadders."

I will continue to use the term "conspiracy theorist" (or the shorter "conspiracist") because the dictionary definition and common usage of it very accurately describe many of the people that I have encountered online, that I have interviewed, and that I have met in person. They are in fact people who *tend* to believe in conspiracy theories as explanations for all major events in the world. I do not mean it to be derogatory, and indeed I would point out the many positive associations in popular culture. I use the term because people understand what it means.

CHAPTER TWO

Conspiracy Spectrums

If you want to understand how people fall for conspiracy theories, and if you want to help them, then you have to understand the conspiracy universe. More specifically, you need to know where their favorite theories are on the broader spectrum of conspiracies.

What type of person falls for conspiracy theories? What type of person would think that the World Trade Center was a controlled demolition, or that planes are secretly spraying chemicals to modify the climate, or that nobody died at Sandy Hook, or that the Earth is flat? Are these people crazy? Are they just incredibly gullible? Are they young and impressionable? No, in fact the range of people who believe in conspiracy theories is simply a random slice of the general population.

Many dismiss conspiracy theorists as a bunch of crazy people, or a bunch of stupid people, or a bunch of crazy stupid people. Yet in many ways the belief in a conspiracy theory is as American as apple pie, and like apple pie it comes in all kinds of varieties, and all kinds of normal people like to consume it.

My neighbor down the road is a conspiracy theorist. Yet he's also an engineer, retired after a successful career. I've had dinner at his house, and yet he's a believer in Chemtrails, and I'm a Chemtrail debunker. It's odd; he even told me after a few glasses of wine that he thinks I'm being paid to debunk Chemtrails. He thought this because he googled my name and found some pages that said I was a paid shill. Since he's a conspiracy theorist he tends to trust conspiracy sources more than mainstream sources, so he went with that.

I've met all kinds of conspiracy theorists. At a Chemtrails convention I attended there was pretty much the full spectrum. There were sensible and intelligent older people who had discovered their conspiracy anything from a few months ago to several decades ago. There were highly eccentric people of all ages, including one old gentleman with a pyramid attached to his bike. There were people who channeled aliens, and there were people who were angry that the alien-channeling people were allowed in. There were young people itching for a revolution. There were well-read intellectuals who thought there was a subtle system of persuasion going on in the evening news, and there were people who genuinely thought they were living in a computer simulation.

There's such a wide spectrum of people who believe in conspiracy theories because the spectrum of conspiracy theories itself is very wide. There's a conspiracy theory for everyone, and hence very few people are immune.

The Mainstream and the Fringe

One unfortunate problem with the term "conspiracy theory" is that it paints with a broad brush. It's tempting to simply divide people up into "conspiracy theorists" and "regular people"—to have tinfoil-hat-wearing paranoids on one side and sensible folk on the other. But the reality is that *we are all conspiracy theorists*, one way or another. We all know that conspiracies exist; we all suspect people in power of being involved in many kinds of conspiracies, even if it's only something as banal as accepting campaign contributions to vote a certain way on certain types of legislation.

It's also tempting to simply label conspiracy theories as either "mainstream" or "fringe." Journalist Paul Musgrave referenced this dichotomy when he wrote in the *Washington Post*:

> *Less than two months into the administration, the danger is no longer that Trump will make conspiracy thinking mainstream. That has already come to pass.*[1]

Musgrave obviously does not mean that shape-shifting lizard overlords have become mainstream. Nor does he mean that Flat Earth, Chemtrails, or even 9/11 Truth are mainstream. What he's really talking about is a fairly small shift in a dividing line on the conspiracy spectrum. Most fringe conspiracy theories remain fringe, most mainstream theories remain mainstream. But, Musgrave argues, there's been a shift that's allowed the bottom part of the fringe to enter into the mainstream. Obama being a Kenyan was thought by many to be a silly conspiracy theory, something on the fringe. But if the president of the United States (Trump) keeps bringing it up, then it moves more towards the mainstream.

Both conspiracy theories and conspiracy theorists exist on a spectrum. If we are to communicate effectively with a conspiracy-minded friend we need to get some perspective on the full range of that spectrum, and where our friend's personal blend of theories fit into it.

There are several ways we can classify a conspiracy theory: how scientific is it? How many people believe in it? How plausible is? But one I'm going use is a somewhat subjective measure of how *extreme* the theory is. I'm going to rank them from 1 to 10, with 1 being entirely mainstream to 10 being the most obscure extreme fringe theory you can fathom.

This extremeness spectrum is not simply a spectrum of reasonableness or scientific plausibility. Being extreme is being on the fringe, and fringe simply denotes the fact that it's an unusual interpretation and is restricted to a small number of people. A belief in religious supernatural occurrences (like miracles) is a scientifically implausible belief, and yet it is not considered particularly fringe.

Let's start with a simple list of actual conspiracy theories. These are ranked by extremeness in their most typical manifestation, but in reality, the following represent topics that can span several points on the scale, or even the entire scale.

1. Big Pharma: The theory that pharmaceutical companies conspire to maximize profit by selling drugs that people do not actually need
2. Global Warming Hoax: The theory that climate change is not caused by man-made carbon emissions, and that there's some other motive for claiming this
3. JFK: The theory that people in addition to Lee Harvey Oswald were involved in the assassination of John F. Kennedy
4. 9/11 Inside Job: The theory that the events of 9/11 were arranged by elements within the US government
5. Chemtrails: The theory that the trails left behind aircraft are part of a secret spraying program

6.   False Flag Shootings: The theory that shootings like Sandy Hook and Las Vegas either never happened or were arranged by people in power

7.   Moon Landing Hoax: The theory that the Moon landings were faked in a movie studio

8.   UFO Cover-Up: The theory that the US government has contact with aliens or crashed alien crafts and is keeping it secret

9.   Flat Earth: The theory that the Earth is flat, but governments, business, and scientists all pretend it is a globe

10. Reptile Overlords: The theory that the ruling classes are a race of shape-shifting trans-dimensional reptiles

If your friend subscribes to one of these theories you should not assume they believe in the most extreme version. They could be anywhere within a range. The categories are both rough and complex, and while some are quite narrow and specific, others encapsulate a wide range of variants of the theory that might go nearly all the way from a 1 to a 10. The position on the fringe conspiracy spectrum instead gives us a rough reference point for the center of the *extent* of the conspiracy belief.

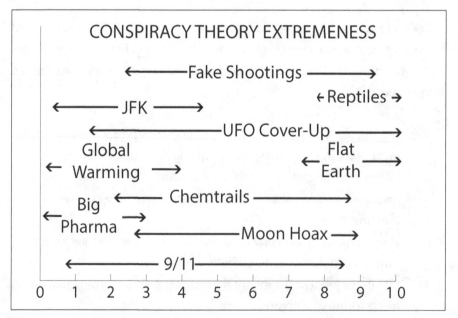

FIGURE 3: A rough overview of the conspiracy spectrum.

Figure 3 is an illustration (again, somewhat subjective) of the extents of extremeness of the conspiracy theories listed. For some of them the ranges are quite small. Flat Earth and Reptile Overlords are examples of theories that exist only at the far end of the spectrum. It's simply impossible to have a sensible version of the Flat Earth theory due to the fact that the Earth is actually round.

Similarly, there exist theories at the lower end of the spectrum that are fairly narrow in scope. A plot by pharmaceutical companies to maximize profits is hard (but not impossible) to make into a more extreme version.

Other theories are broader in scope. The 9/11 Inside Job theory is the classic example where the various theories go all the way from "they lowered their guard to allow some attack to happen," to "the planes were holograms; the towers were demolished with nuclear bombs." The Chemtrail theory also has a wide range, from "additives to the fuel are making contrails last longer" to "nano-machines are being sprayed to decimate the population."

There's also overlapping relationships between the theories. Chemtrails might be spraying poison to help big pharma sell more drugs. JFK might have been killed because he was going to reveal that UFOs were real. Fake shootings might have been arranged to distract people from any of the other theories. The conspiracy theory spectrum is continuous and multi-dimensional.

Don't immediately pigeonhole your friend if they express some skepticism about some aspect of the broader theories. For example, having some doubts about a few pieces from a Moon-landing video does not necessarily mean that they think we never went to the Moon, it could just mean that they think a few bits of the footage were mocked up for propaganda purposes. Likewise, if they say we should question the events of 9/11, it does not necessarily mean that they think the Twin Towers were destroyed with explosives, it could just mean they think elements within the CIA helped the hijackers somehow.

Understanding where your friend is on the conspiracy spectrum is not about which topics he is interested in, it's about where he draws the line.

The Demarcation Line

While conspiracy theorists might individually focus on one particular theory, like 9/11 or Chemtrails, it's very rare to find someone who only believes in one conspiracy theory. They generally believe in *every* conspiracy theory that's less extreme than their favorite one.

In practical terms this means that if someone believes in the Chemtrail theory they will also believe that 9/11 was an inside job involving controlled demolition, that Lee Harvey Oswald was just one of several gunmen, and that global warming is a big scam.

The general conspiracy spectrum is complex, with individual theory categories spread out in multiple ways. But for your friend, an individual, they have an internal version of this scale, one that is much less complex. For the individual the conspiracy spectrum breaks down into two sets of beliefs—the reasonable and the ridiculous. Conspiracists, especially those who have been doing it for a while, make increasingly precise distinctions about where they draw the line.

The drawing of such dividing lines is called "demarcation." In philosophy there's a classical problem called the "demarcation problem," which is basically where you draw the line between science and non-science. Conspiracists have a demarcation line on their own personal version of the conspiracy spectrum. On one side of the line there's science and reasonable theories they feel are probably correct. On the other side of the line there's non-science, gibberish, propaganda, lies, and disinformation.

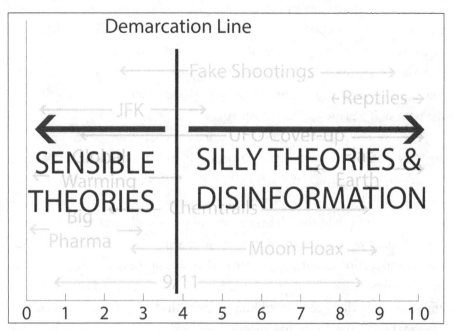

FIGURE 4: The line of demarcation. Everything below it is sensible, everything above it is silly. Where do you draw the line?

I have a line of demarcation (probably around 1.5), you have one, your friend has a line. We all draw the line in different places. But we all have remarkably similar assessments of the things on the more extreme side of the line. Something that often surprises me is how often people with low-level theories (like the simpler 9/11 theories) tell me that they like my site because they use it to help debunk Chemtrails or Flat Earth. They are even happy when I debunk things that are just on the other side of their line, but within their general conspiracy theory (like 9/11). Even people who seem convinced I'm some kind of government agent have said this, like Ken Doc, the organizer of one of the more sizable 9/11 Facebook groups:

> *[Mick] West is a shill for the Government and believes there is no such thing as "Conspiracy Theories" or "Government Corruption." He is great at debunking the disinfo alternative theories in the Truth Movement, but when it comes down to the science and physics of 9/11, he suffers from cognitive dissonance.*[2]

This quote illustrates two issues of demarcation. Firstly, Ken has a very clear line in his mind where on one side there is "science and physics" which he thinks prove controlled demolition, and on the other side there are "disinfo alternative theories," which is anything that he thinks goes a bit too far. Ken for example draws the line at an event that divides the 9/11 community: the issue of if a plane hit the Pentagon.

Secondly, issues that are close to the line are contentious, even those that are only just on the other side. Many people in the 9/11 community think that it's just a simple and obvious matter of science and physics that the Pentagon was hit by a missile. They will point to the size of the hole, or the lack of plane parts in the debris. These people obviously all think that the World Trade Center buildings were destroyed with explosives, just like Ken does. Ken's demarcation line lies just on the other side of the issue. He thinks that the "Pentagon missile" theory is "disinfo," something that's obviously wrong, and something that needs debunking along with the even more extreme 9/11 theories like "nukes in the basement," "energy beams from space," or "all the footage was CGI." He's quite happy for me to focus on things on that side of his personal demarcation line.

Ken has an extensive website in which he makes a very detailed case for his version of the 9/11 inside job, controlled demolition theory.[3] But on that site the line is clearly defined in his section on "9/11 disinfo":

Disinformation is false information that is used to dismiss legitimate arguments. Posted in this thread are many examples of Disinformation and/or speculative theories that have been purposely spread in order to divide, mislead, and/or to discredit the 9/11 Truth Movement.

We all have our lines, and our perception of the claims on the more extreme side of the lines is that it's just *false information*. We might disagree about the reasons for that false information existing, but we all think the stuff on the extreme side of the line is wrong. Helping people out of the rabbit hole can equate to simply moving that line gradually down the extremeness spectrum. But to move their line you've got to understand exactly where that line of demarcation lies.

The demarcation problem occurs up and down the conspiracy spectrum, and the conspiracists all feel the same way about it. There's the sensible side, and there's the silly side. This brings a significant problem in communication—nobody wants to be associated with the silly side. They often think that anyone who brings up the "silly" side is deliberately trying to discredit them. Ken Doc is a 9/11 Truther and resents any kind of association with Chemtrails. JFK theorists resent being compared to Truthers; Chemtrail believers resent being compared to Flat Earth believers.

One obscure branch of the 9/11 Truth community is the "no-planers," a group who thinks that the all the television footage was faked, and that nothing hit the towers. This theory does not stand up to the slightest scrutiny, as there were tens of thousands of eyewitnesses in Manhattan and the surrounding area. But there's still a lively discussion forum (the Clues Forum) where they swap ideas and bits of "evidence" they have gleaned from their research.

Being so extreme means they also embrace other high-order conspiracies, like Chemtrails and the Moon-landing hoax. But apparently the point where many of them draw the line is with the Flat Earth theory, which they consider to be "discrediting by association" (DBA) and a deliberate falsehood spread by NASA. This is explained by "simonshack" on the Clues Forum.

> *What NASA has rolled out is a carefully planned and coordinated, 'viral' DBA/ co-opting campaign centered on the 'Flat Earth' meme and—I will hastily add—(with respect to those who may honestly entertain alternative cosmic models of their own) this fact should be clear as day to anyone, regardless of whether you reckon we live on a globe, a cube, a pancake or a Wiener würstel. The point being, it is by now painfully obvious that the objective of this NASA-campaign is to associate & equate (in the general public's psyche) anyone questioning NASA to mentally challenged raving crackpots.*

Here we've got a group of people who think that the entire space program is fake. They think we never went to the Moon, and the International Space Station footage is filmed in a studio somewhere. Yet they think that the Flat Earth theory goes too far, and so must have been invented to discredit them.

They even go a little beyond that. Notice how simonshack says, "regardless of whether you reckon we live on a globe." He's not throwing the baby out with the bathwater here—it's not specifically the Flat Earth theory that he thinks is disinformation, it's *the crazy version* of the Flat Earth theory that's the NASA creation. He draws his line of demarcation to encompass sensible inquiry into the possibility that the Earth might not be a globe.

Your friend will have a line of demarcation. This is both something you need to be careful of, and something that you can use as an illustrative tool. Firstly, be very aware of this problem of "discrediting by association." Be clear that you are not trying to lump them in with people on the other side of their line. Tell them (honestly) that it's good that they haven't been sucked deeper in. Don't stray *too* far into comparisons with the other side of their line, as they might find things like "at least you're not a Flat Earther!" to be poking fun at them. Instead focus on the aspects of their belief that are very close to the line. If they are a 9/11 Truther this might be the issue of what happened at the Pentagon. If they are a Chemtrail believer, it might be the question of if contrails can persist naturally.

Your awareness of where they draw the line can be used to defuse an argument from authority. Many people base their beliefs in large part on what they see as information from trusted sources. With 9/11 this is frequently people like Architects and Engineers for 9/11 Truth. If you can show them that their trusted source believes in something that's on the other side of *their* line, then it can be something that opens them up to the idea that perhaps the things that that source has been telling them might not be as reliable as they thought. My friend, Steve, explains how this happened for him.

> *The We Are Change people [a 9/11 Truth group] were very seriously into Chem-*
> *trails. For me it was just a simple thing, looking up the word "contrail" . . . it*
> *said there were ones that were short, and others persisted and spread out, and*
> *the reasons for that. So, I tried to explain that to the people at We Are Change,*
> *and they threw me out of We Are Change temporarily. So that kind of under-*
> *mined the whole cause of 9/11 Truth.*

The realization that their trusted source actually is on the other side of their line can happen just by finding out something about the source (for example, you might find that your favorite 9/11 conspiracy theory writer also thinks that the Sandy Hook children were not killed). It can also be something that you help happen by moving the line a little by debunking one of the claims that a source makes. For example, if your friend's source says contrails can't persist, and you show them conclusive proof that they can (by showing them decades of books on the topic), then you've moved the line, and raised doubts about the sources of their information.

It's worth bring up a note of caution yet again here: *be honest and open.* If your friend is a conspiracy theorist, then they are going to be suspicious. One thing they are going to be suspicious of is you, and the tactics that you use. If they think you are trying to contrive an argument simply to make their source look bad, then this can have a significant backfire effect. They might accuse you of running a smear campaign, attacking the person instead of the evidence. Make sure that you are being honest from the outset, and that you only use verifiable facts.

Taxonomies of Conspiracies

Besides a simple measure of "extremeness," there are other useful ways we can categorize conspiracy theories. In his book, *A Culture of Conspiracy,* Michael Barkun says they can be divided into three types: Event Conspiracies, Systemic Conspiracies, and Super Conspiracies. Event Conspiracies are those that focus around a single event, such as the assassination of John F. Kennedy, or the terrorist attacks of 9/11. Systemic Conspiracies are those that involve complex plots that continue over a long period of time, such as water fluoridation or Chemtrails. Super Conspiracies consist of multiple separate conspiracies spanning the entire spectrum of subjects, all linked together into one overarching master plan.[4]

Of these, the most common is the Event Conspiracy. Event Conspiracies now spring up immediately after almost any event that makes the evening news, even for seemingly mundane happenings. When the lights went out in the stadium at the 2013 Super Bowl, there were almost immediate suggestions that it was done deliberately, perhaps by some shadowy hacker group such as Anonymous, in order to give the trailing San Francisco 49ers a chance to regroup. This was probably reinforced in the minds of the suspicious when the 49ers managed to rally in the second half, almost defeating the Baltimore Ravens.

Much more disturbingly, Event Conspiracy theories sprang up immediately after the shootings of twenty children and six adults at the Sandy Hook Elementary School in Newtown, Connecticut, and in other shootings and bombings. There's a huge range of Event Conspiracy theories. We could group them into one of four sub types, in increasing order of improbability:

EXPLOITED EVENT—THE "GLAD IT HAPPENED" THEORY. Here the event is genuine, and the "conspirers" are as surprised as anyone that it happened—however they immediately begin to exploit that event, and spin, lie, and distort what actually happened to further their goals. Here 9/11 was supposedly re-framed in such a way that many people got the impression that Saddam Hussein was responsible for the attacks, and this was used to provide justification for the Iraq War. The Sandy Hook shootings were supposedly misreported and exploited in order to promote gun control. These theories are often quite plausible.

ALLOWED EVENT—THE "LET IT HAPPEN" THEORY. Here the events are as they appeared to be. The 9/11 attacks were performed by terrorists hijacking planes. JFK was shot by Lee Harvey Oswald. But in this scenario, there is a set of secret conspirers (usually people in power, such as the executive branch of the government, or elements of government agencies) who are supposedly aware of the planned event ahead of time and could have stopped it by warning people. But they kept silent, and let the event happen because it benefits them in some way. Here George W. Bush supposedly allowed the attacks on 9/11 because they would provide justification for invading Iraq. Here the attacks on Pearl Harbor were supposedly known days in advance by the US and/or British government, but they let it happen to provide popular support for the US entering the Second World War.

DELIBERATE EVENT—THE "MADE IT HAPPEN" THEORY. Here the events are real, but they were performed or ordered by the people behind the conspiracy. In this scenario the World Trade Center was hit by remote control planes and the buildings brought down by controlled demolition. JFK was supposedly shot by a CIA sniper. The Sandy Hook children were supposedly shot by gunmen dressed as nuns who left Adam Lanza's drugged and shot body at the school after they had slaughtered the innocents.

FAKED EVENT—THE "IT DIDN'T HAPPEN" THEORY. Here the entire event is a concoction of the government and the media. In this worldview, we are living in a constructed reality. This is not to say we are living in a computer simulation, but that nearly everything portrayed in the media has been faked. Supposedly nobody landed on the Moon, planes never struck the World Trade Center, children weren't killed at Sandy Hook. These events were all somehow staged to provide justification for some action. The footage of 9/11 planes were computer animated, the videos are all faked, the people running in the street were all actors.

While you can usually describe any particular event conspiracy theory by labeling it as one of the above four types, it's quite rare that it fits neatly within a single category. In particular the exploitation of an event is presumed to be happening regardless of if the event was allowed, deliberate, or (especially) faked. Some events are supposedly halfway between "Allowed" and "Deliberate," perhaps Adam Lanza (the Sandy Hook shooter) was known to be a psychopath with access to guns, and was supposedly given drugs that made him violent, or perhaps he was brainwashed by voices beamed into his head at night. A kind of "helped it happen" theory.

It's important when talking to your friend to be clear which particular type of conspiracy you are dealing with. Most people would agree that elements within the Bush administration exploited the events of 9/11. Fewer, but still a significant number, think that Bush had some kind of foreknowledge of some kind of attack, at least to the extent of not fully acting on warnings. Fewer still think that Bush knew specifically what the attacks would be. A very small percentage of people think that Bush and others in power deliberately engineered the attacks using the terrorists. A smaller percentage thinks that the attacks happened, but not with terrorists, but with remote control planes, and deliberate controlled demolition. A vanishingly small number of people think that the attacks did not happen at all and were entirely faked.

And yet the people who promote the more exotic conspiracy theories (like controlled demolition of the World Trade Center) will often use the number of people who suspect the more mundane (like Bush exploiting the events to invade Iraq) to bolster their case. They will take the fact that a large number of people are naturally suspicious of the actions of the government relating to 9/11 and try to hijack those numbers to suggest that their own bizarre theory of controlled demolition has broad popular support. But while many people would sign a petition that called for a new and open investigation of the events of 9/11, not all of those people, and probably not even a significant fraction, strongly believe that the buildings were deliberately destroyed by Bush administration operatives with explosives.

The 9/11 Truthers have their own internal taxonomies, generally referred to as IHOP. IHOP stands for It Happened On Purpose. The variations on this taxonomy relate to the degree of intentionality behind an event—a more nuanced version of the event conspiracy scale discussed previously. The acronyms LIHOP and MIHOP were originally invented to describe a division at the less extreme end of the 9/11 conspiracy theory spectrum. Did the Bush Administration let it happen on purpose (LIHOP), or did they actually proactively *make* it happen (MIHOP)?

Nicholas Levis, writing for the now defunct 9/11 Truther website Summer of Truth, devised a ten-point "HOP" scale.[5] The first five graduations are degrees of let-it-happen, ranging from a kind of benign lack of preparedness to a "LIHOP PLUS" at level 5, where the Bush administration helped bin Laden, at least by moving problems out of the way. Levis himself subscribed to this LIHOP PLUS theory.

Levis's MIHOP scale starts with what we generally think of as the most common 9/11 Truth theory—that it was all planned by the US Government, that the buildings were destroyed by remotely controlled pre-planted explosives, and the planes were possibly flown by remote control. The scale after that is not so much an increasing level of MIHOP-ness, but more a variety of different scenarios for who was responsible—the New World Order, rogue Neocons, or a rogue state such as Israel, China, or Russia.

The LIHOP crowd are not particularly active. Most 9/11 Truthers seem to fall within the MIHOP PLUS category. Within that they have various other divisions, like "planers/no planers" or "thermite/DEW/nukes" (for the various proposed method of destruction).

A key part of helping your friend is figuring out where they sit within these complex taxonomies, so you can effectively talk to them. What's their favorite conspiracy? Where do they draw the line within that conspiracy? How do they feel about other conspiracies? What, to them, are reasonable theories, and what are unscientific nonsense and disinformation?

As well as understanding your friend, you want to make sure that they understand you. Partly that's just about sharing what you know about the topic, and about the context surrounding the topic. But sometimes it's also about demonstrating that you are genuine in your interest, that you are not a "useful idiot," and that you are not a shill.

CHAPTER THREE

The Shill Card

The "shill card" is a tactic used by the promoters of disinformation to discredit people who point out their errors. Instead of answering the objections being raised, they simply claim that the source is a "government shill." It's an old tactic, more generally described as "poisoning the well." It is applied to fact checking sites like Snopes or FactCheck.org, it's applied to the media by politicians, and it's applied to debunkers by conspiracy theorists.

Once a label has been applied it often sticks. Posting an explanation for something like contrails with a reference to Wikipedia might be met with a response like "Wikipedia, LOL," or even "Wikipedia has been debunked." A very useful source of largely neutral information has been suddenly deemed entirely unreliable. Not only that, but the fact that Wikipedia (or Snopes, or Metabunk) says something is often interpreted (by conspiracy theorists and people on the political fringes) as evidence that the opposite is true.

If you encounter this accusation after referencing a site like Wikipedia, then the best thing to do is simply bypass that "tainted" site and use the direct references instead (Wikipedia in particular almost always has the information available in several referenced articles). But what do you do when the shill card is played against you yourself?

This is a position I have found myself in many times, and I have also seen it when other people reference articles I've written. Like the "Wikipedia, LOL" comeback, I get similar responses, like this one from a believer in the "Flat Earth" theory:

Anyone who buys Metabunk bullshit deserves not the truth. Mick West, lol, what a joke. He is utterly ignorant of pretty much everything involving basic physics.[1]

Or from a believer in Chemtrails:

When a liar is caught in a lie, the only way to cover up the previous lie is to make up another lie. That is why Mick West, the most notorious internet troll in existence, will never tell the truth.[2]

Or from a believer in the 9/11 Controlled Demolition theory:

Mick West is a shill for the Government and believes there is no such thing as "Conspiracy Theories" or "Government Corruption." . . . When it comes down to the science and physics of 9/11, he suffers from cognitive dissonance.[3]

I've even seen it in strange indirect manners when people on social media raise issues that sound like claims I've made (like, "contrails have always persisted"), then they get accused of being Mick West in disguise, and their evidence is discarded.

I have tried to counter this by being open, honest, and respectful. I tell people who I am. I explain to them what I think. I explain that I think there are many real conspiracies, an abundance of government corruption, and that nobody should blindly trust people in power. I discuss why I debunk *false* conspiracy theories, and why I think that is important. I discuss my past, explaining how I got here, what my credentials are (more often how I don't have any), how I came to know so much about contrails, and how I can afford to run Metabunk by myself.

The short version I give is that I'm a retired video game programmer and nobody pays me. I debunk as a hobby and it's something I've always been interested in. Metabunk is just an off-the-shelf forum that costs very little to run except for time, of which I have plenty.

The longer version follows. I include my story here because the more someone knows about you and the more they can relate to you, the more effective your communication will be. That does not necessarily mean they are more likely to believe in what you say, but it will at least *help* get past the "shill card." If I can get across to people the fact that I don't need anyone to pay me, and that I genuinely think that debunking false conspiracies is a valuable thing, then they will (hopefully) move away from the position that I am being paid to lie, and towards the belief that I am simply mistaken, that my disagreement with them is honest. And from that position we can have a much more productive discussion.

I was born in the 1960s in Bingley, a small town in the north of England. My parents, two sisters, one brother and I were all packed into an old stone terrace house originally built to house workers from the nearby wool mill. We were a poor family and did not have a phone or a television for many years. I learned to read with my father's collection of Marvel comics, and later his large collection of science fiction. I was an average student who read a lot, but I excelled in mathematics. I really enjoyed solving math problems, particularly those to do with physics—calculating things like velocity, acceleration, energy, and momentum.

My grandfather encouraged my interest in mathematics; he bought me a programmable Casio calculator and asked me to try to program it. I discovered that not only was I merely capable, in fact I also really enjoyed doing it. I spent a cold winter delivering newspapers to raise enough money to buy my first computer and began to learn how to program. I became obsessed with programming and playing computer games (which were very simple things back then).

Another obsession of mine was reading. I read a lot of science fiction, but I also read a magazine called *The Unexplained: Mysteries of Mind, Space and Time*, a periodical published in the early 1980s that told (supposedly true) tales of UFOs, ghosts, magic, and other strange beasts that people were unable to explain.

For many years these things were causes of great fear to me. In my early teens I used to lie awake at night, literally trembling with the thought that some alien could enter my room and spirit me away to perform experiments on me, or that ghosts might actually be hovering around me, ready to shriek into existence, or softly stroke my face with disembodied hands reaching out of the darkness.

In particular one small book, written for children, really scared me. It contained an account of the Kelly–Hopkinsville Encounter, a "true story" of a farmhouse under attack by little green men. At one point they describe turning around to see a clawed hand reaching down towards them.

But as I grew up and learned more of science, and the way the world actually worked, these fears dropped away. I found that the Kelly–Hopkinsville "aliens" were almost certainly owls, and my fears felt silly in hindsight. I did not lose my fascination with these fringe topics, but instead became even more interested in them, particularly in seeking out the most reasonable explanations of strange phenomena. I never fully rid myself of the fear—I'm still not entirely comfortable in the dark. I can rationalize it away, I know the fear is an illusion, but it's still there.

A small part of the reason why I debunk now (and still occasionally address ghost stories) is anger at the fear this nonsense instilled in me as a young child. Perhaps you can't do much to stop children being afraid of the dark, but I can still call out the bunk in the tales, and perhaps that will help someone be less afraid. Perhaps it will stop people from passing off these stories as true. Every little bit helps.

Progressing through school I continued to do well in math and physics. I also studied advanced level draftsmanship, following the profession of my grandfather. I went on to university in nearby Manchester to study computer science. I spent far too much time playing with my computer and reading science fiction, but I managed to scrape by with a degree.

In my last year at university I entered a national contest to describe the future of information technology. Skipping classes for a week, I created a futuristic newspaper describing something called "The Stream," which quite closely resembled the present internet. I won the contest[4] and a cash prize which allowed me to pay a few months' rent on an apartment, and to get a modem.

There was no internet at the time, but there were a few very small BBSs (bulletin board systems) which typically only one person could use at a time. Connection speeds were measured in bits per second, literally a million times slower than today's fastest connections. The closest thing to a public internet was FidoNet, a collection of modem-based bulletin boards that called each other up at night to swap information. Interactions were necessarily slow, so two people would often exchange only one message a day, or less.

But it was with this limited online presence that first I took up debunking as a hobby. I continued to read the *Unexplained* magazines, but now the greatest source of fascination to me was the explanations. Spontaneous Human Combustion became less a source of fear that I might suddenly burst into flames, and more a macabre wonderment that a human corpse could burn, the flames fed by body fat, a good source of air, and the wicking action of clothes. I liked to share this information with my friends, or with anyone who thought Spontaneous Human Combustion might be supernatural.

My competition winnings ran out in a few months and I was forced to get a job. Luckily my skill at math, my love of computers, and my ability to solve problems made me an ideal candidate to work in video games. I was in precisely the right place at the right time with the right set of skills. Back then games were much easier to create, and often the programmers had no formal training, having entered the industry in their teens. I got a job writing a snooker (billiards/pool) game and began my career as a video game programmer.

I moved to Los Angeles, California, in 1993, where I worked for a year at Malibu Interactive, writing a robot war game. Once more I found myself in the right place at the right time, during a period of rapid expansion of the industry, especially in Los Angeles. I left Malibu with Joel Jewett and Chris Ward to start our own company, which (somewhat on a dare) we called Neversoft. It was touch-and-go for a few years, but we eventually hit our groove with the wildly successful *Tony Hawk's Pro Skater* series.

I was mostly busy with work for the next decade, consequently much of my debunking was at work—as emails would get passed around (after we actually got email, around 1996), I was always quick to point out the errors, and direct the writer to something like Snopes (which was founded in 1995). I remember one particular story around the time of the Mad Cow scare (around 1995), where a brain disease infecting cows was occasionally being transmitted to humans as Creutzfeldt–Jakob Disease (CJD). It was claimed that 50 percent of Britons (including myself) would die of CJD disease within ten years. That was over twenty years ago and 99.9999 percent of Britons are entirely unaffected, but the effects of that media scare live on, in that I *still* can't donate blood in the US.[5] This always stayed with me as a prime example of the negative effects of junk science—the predictions of a CJD epidemic in the UK were vastly overblown, and in the twenty years since the fears were raised there has never been a proven case for CJD from blood transfusion.

I took up my hobby of debunking more seriously again after I cashed in my stock options and left Neversoft in 2003. The Tony Hawk money meant I could pretty much retire, giving me a lot of freedom to do whatever I liked with my time. I started part-time work as a writer for *Game Developer* magazine, where I just wrote about whatever interested me in game development technology. Around 2005 I joined Wikipedia as an editor (that does not mean I worked there; anyone can join). Initially I did lots of minor little editing on dubious topics like homeopathy and audiophiles. I found a big source of bunk in the form of Biblical Scientific Foreknowledge (now called "Scientific foreknowledge in sacred texts"), which is rather a fringe subject that suggests that there is scientific truth in the Bible that was not available to humanity at the time it was written, the claim being that this was proof that the Bible was written by God. I delved into such arcane subjects as Ancient Egyptian Medicine, biblical exegesis, phytopharmacology, and vegetarian lions.

At some point on Wikipedia I found an article on a proposed medical condition called "Morgellons." It looked a bit dubious to me. People were claiming that fibers were coming out of their skin, and I thought that looked like they might just be mistaking random bits of clothing fibers. I made some edits to the page in March and April 2006. After a few weeks, I found the subject so interesting that I started my first single-subject skeptical blog (using the blogger.com platform at the time, later switching to WordPress, both of which were free).

MorgellonsWatch.com was my part-time hobby for nearly three years. I wrote over a hundred articles, and got over 12,000 comments. During the first year or so there were several media stories regarding Morgellons, and I received a few requests for interviews. I declined them because I wanted to remain anonymous. I was actually a little embarrassed by the amount of work I put into my odd hobby, and preferred to not discuss it with anyone.

I learned many lessons while running Morgellons Watch. The most important one was to be polite and respectful to people that you disagree with. Nothing good comes of insulting someone, even unintentionally.

Interest in Morgellons died down around 2008 and more or less went away after the CDC did a study on the topic in 2012 and found nothing unusual. So I stopped posting and moved on to other topics.

After I left Neversoft one of the first things I did was take flying lessons at Santa Monica airport. It turned out flying was a bit more stressful than I'd imagined. Santa Monica airport is right next to the very busy Class-B airspace of Los Angeles International Airport, and to go anywhere requires careful planning and often complex navigation. I got my solo certification, I did a few long distance solo flights, then decided flying was really not for me.

But along the way I had to learn a lot about planes, air traffic, and the atmosphere. I also found a new topic that intersected with my new knowledge and my interest in debunking. The topic was Chemtrails—the unfounded belief that the long white trails left by high flying airplanes were not just condensation clouds, but were artificial trails deliberately sprayed for some illegal or nefarious purpose.

I came across the topic of Chemtrails somewhat randomly on Wikipedia, and I immediately found it fascinating—particularly the false idea people had that "normal" contrails could not persist. At that point (in 2007) I'd grown a little disillusioned with Wikipedia. Rather than spend much time editing the Wiki article (which would often be re-edited by a believer in Chemtrails shortly thereafter), I simply started a new blog: Contrailscience.com.

Chemtrails *seemed* like safer ground than Morgellons, as it seemed to mostly be a misunderstanding of the physics of the atmosphere with none of the implications of mental illness that Morgellons had. But there was still this constant problem that people got so upset by rigorous criticism of their ideas that productive conversation became impossible. So to try to address this I instituted a politeness policy on the site which grew more and more strict as time went by.

I was still settling into semi-retirement. After Neversoft I'd done some consulting work. I wrote the artificial intelligence for the computer players in a poker game. I wrote articles on topics like simulating fluid mechanics or analyzing video game lag. I was contracted by a large corporation to build a robot to test the effects of various factors on the response times of video game controllers. I wrote an iPhone app (one of the first available at the launch of the app store) to help Scrabble players. I traveled the world with my wife. I spent a lot of time on the internet—anonymously posting explanations of "Chemtrails" on Contrail Science. But I was getting a little bored of the whole thing, and I considered closing it down and spending more time programming.

Then in December 2009 I had a story hit the big time with the case of the "Mystery Missile"—where a plane flying from Hawaii to the mainland left a contrail on the horizon that looked a bit like a missile trail. A Los Angeles news chopper spotted it, put it on the evening news, and the story went viral. I wrote a few articles debunking this (explaining how it was just a contrail from an odd angle) and ended up being contacted by the media with inter-view requests. At the time I was still anonymous, but I decided then that my debunking would be taken more seriously if I was honest about who I was. This was a pivotal moment. I could either maintain my anonymity and just slip away, or take advantage of this publicity to get my message, one of truth and science, out there. I "broke cover," and went on *CNN* and *CBS Evening News* to explain what the trail actually was.

I got a lot of traffic from that, nearly a million visitors to Contrail Science over the course of a week. This felt like a good time to branch out a bit. There was a lot of discussion in the comments section on Contrail Science, but the blog format was cumbersome, and the topics quite often strayed from the topic of contrails and onto broader conspiracies or other strange phenomena. I decided to set up a forum to foster more of that wider conversation.

Metabunk.org was born in December 2010. The name Metabunk is meant to convey the idea of thinking about debunking, and not simply doing it. Thinking about what bunk is, how to debunk better, and thinking about why we debunk and what it is we are really trying to do. I write longer articles that get published to the front page, but Metabunk is really arranged as a multi-user forum with various sub-forums. Some sub-forums are on the more "meta" topics, like "Practical Debunking," and "Escaping the Rabbit Hole," but there's more activity in the topic-specific forums, like "Contrails and Chemtrails" (where all the Contrail Science discussion now lives), "9/11," and "Flat Earth." There's also a thriving micro-community of people in the "Skydentify" forum who enjoy tracking down odd things in the sky (usually aircraft with contrails).

The Chemtrail theory remains the most popular subject, but I cover a wide range of topics. There's quite a bit of photo analysis, including photos of "ghosts," UFOs, lake monsters, and other things straight out of the pages of *Unexplained* magazine. The 9/11 forum gets a reasonable amount of traffic, but curiously the most popular posts are esoteric things like fireproof snow or Flat Earth.

Since there are not many people writing about Chemtrails in depth, I tend to get calls from the media when they are doing stories about it. One of these calls was for a show called *Joe Rogan Questions Everything*, hosted by Joe Rogan himself, who I only knew as the former host of the *Fear Factor* game show. Joe was someone who used to be more into strange beliefs, like UFOs and conspiracy theories, but was gradually becoming more thoughtful about them after he met a wider variety of both believers and skeptics.

After the episode aired Joe invited me on his very popular podcast, *The Joe Rogan Experience*, to discuss the Chemtrails theory. This was good publicity for Metabunk, but also seemed to convince some people that both Joe and I were government shills. But it also had a positive effect for many people, as I was told via email recently:

I think what was so helpful about your appearances on Mr. Rogan's podcast was that it helped break a common line of thinking I think many conspiracy theorists, and certainly those I knew, suffer from. They have a tendency to think any person who attempts to disprove or debunk their theories must have some ulterior motive (commonly manifested in the tendency to call debunkers shills or suggest that they are "bought and paid for").

My friends were already fans of Mr. Rogan, and therefore believed he was a man of integrity. Because of this, they could let go of the belief that you were only debunking their theories due to being an apparatchik of the global cabal as they

often imply, because they trusted Mr. Rogan would not do such a thing. There-
fore, it allowed them to simply listen to your ideas objectively, rather than using
some concept of you being a shill as an excuse to instantly discount and ignore
your arguments. Being as they were genuinely intelligent people, once they
began to listen to your ideas honestly, they were perhaps unsurprisingly con-
vinced by your logical and fact-based approach.

Over the next couple of years, I continued to do minor "talking head" appearances for TV shows, and a variety of podcast interviews. Then in 2016 the Flat Earth theory became quite popular. Joe had a friend who had got sucked into that particular rabbit hole, and he invited me back on his podcast to talk about that. This prompted a temporary shift in focus on Metabunk as I wrote a bunch of articles on Flat Earth to make sure I was prepared. Metabunk briefly turned into the hub of Flat Earth debunking.

The Flat Earth show was shorter than normal as Joe was sick, but the episode still got quite a bit of reaction. I was a called a "globetard shill." Joe did not help here, actually having T-shirts with "Globe Earth Shill" printed on them, which I declined to wear. But there was also some positive feedback, which brings us to where I am now, typing this book. The publisher saw me on the podcast and thought it would make a good book, so they approached me, and here it is.

After all this talk of me being a shill, when this book hits the shelves it will be *the first money I have ever made from debunking*. I've never (other than very brief one-day experiments) had ads on Contrail Science or Metabunk. Nobody has ever paid me. I operate my sites out of my own pocket, and it only costs around fifty bucks a month.

I'm not a shill. I'm just someone who thinks the truth is important, and someone who enjoys finding it and helping other people find it too.

That's my story, told as honestly as I can in the space available. It is laid out here as an answer to the shill card. People will continue to claim that I am a shill, because the primary evidence that they use for that claim is simply that I disagree with them. But by laying out my past, my source of income, my personal history with the topic, and by explaining why I do what I do, I hope to at the very least get some people to consider the possibility that I am not a shill. I'm simply someone who believes in a different version of reality to them. Hopefully they will then ask *why* I have opted for this version of reality where the Earth is round, planes leave harmless clouds, the Twin Towers fell due to fires and gravity, and there is not a cabal of evil bankers controlling every aspect of the world. If they can see that I am an honest man with honest beliefs, then I can finally get to explain to them the reasons for my beliefs.

How does all this apply to you and your friend? In two ways. Firstly, my life story as a debunker is laid out in this chapter so you can avoid being tarred with the same brush. Some of the new information you are going to be presenting to your friend (especially regarding Chemtrails) might come from one of my websites. You might even, in the spirit of openness, tell your friend you have read this book, or even try to get them to read it. If you can show them evidence that I'm not actually a paid shill, then it might make it easier for you.

Secondly, I encourage you to try to be equally open with your friend. You don't need to detail your childhood nightmares, but you can still explain why you personally think it's important that we try to debunk claims that are false. It really hinders communication when people even slightly suspect you might be a shill, so make an honest effort to explain why you are disagreeing with them.

CHAPTER FOUR

The Rabbit Hole: How and Why

In the twenty-first century the most common first step in the journey down the conspiracy theory rabbit hole is watching a video. The reasons why people watch the video in the first place vary. It might just be random chance, or hearing about the theory from a friend, or seeing some discussion online. Intrigued, and initially quite skeptical, they decide to do a little research, so they look something up online, or they watch a video that their friend recommended for them. Today it's largely YouTube videos, but for older theorists it might have been from a shared DVD. That's what happened to Copenhagen professor of chemistry Niels Harrit:

Interviewer: Tell me how you decided to take on this, one of the most controversial issues of our time, the events of September 11.

Harrit: This more or less is a torch which somebody gives you, destiny, and you get involved. My story is not very different from millions of other people, because on the day, and the day afterwards, I didn't think much about it. I, maybe I accepted that there were terrorists who could hijack airlines and they smash into the Twin Towers, etcetera. But it wasn't before 2006 when I accidentally saw a DVD on Building 7, and this was shocking for a natural scientist, because I simply couldn't understand what I saw.[1]

Similar tales are told by many former (and current) believers. For example, Martin Beard, once a highly active member of the UK "Chemtrail" scene:

My story goes back to 2007 when (in my late twenties) I was a single bloke still in pub mode and getting pretty sodden [drunk] each weekend (and most nights) just because I could. Whilst working for one of the biggest Pharma companies in the world, Eli Lilly & Company, I was introduced to big bearded South African chap who asked me to take a look at the Zeitgeist movie . . . That's where it all began . . .

From there I was amazed at what I thought was absolute proof the world we live in was a complete lie. It snowballed from there, Alex Jones, David Icke, Edge TV, The UK Column. . . . My whole world changed in a few months and I very quickly became a recluse, spending more and more time sitting in my then flat and drinking lots to numb the pain of this so-called truth.[2]

A common factor seems to be having some spare time. These are not short videos you might just happen to watch during your lunch break. *Zeitgeist*, (the movie that brought Beard, and many others, down the rabbit hole) is over two hours long. *Zeitgeist* is a rather odd movie that starts out with some speculation about Jesus before delving into 9/11 controlled demolition theories, Jewish world bankers, and the forced implanting of microchips into people to enslave them. The DVD that Harrit saw was probably the more conventional 9/11 conspiracy video *Loose Change*, which runs up to 130 minutes, depending on which version you choose. Once they start, people often binge-watch multiple conspiracy videos. Sometimes they watch the same video over and over again. You get the sense from talking to them that it's something like a drug, that the "truth" they feel is in the video is activating some kind of function in their brain, resonating with them, and fulfilling some kind of need.

There are certain personality and psychological aspects that correlate with a tendency to believe in unfounded conspiracy theories. There's the need to feel unique, an overdeveloped tendency to find patterns in things, there's factors like openness to new ideas, agreeableness, intelligence, and attitudes about authority. Researchers have found some rough correlations between various factors like these and conspiracies.

But these factors are things that everyone has to a varying degree; the correlations are also generally quite small—having one or more of the personality factors simply raises the probability that any random person might be a conspiracy theorist. It does not mean these are the actual causes of conspiracy thinking.

Beard did not become a conspiracy theorist because of some need for uniqueness or because his favorite political party had lost the election. He became a conspiracy theorist because he watched *Zeitgeist* at a time in his life when he was alone and getting drunk every night. While there may well have been personality factors that made his descent into the rabbit hole both more probable and ultimately quicker and deeper, it's also very likely that had he not seen *Zeitgeist* he would have remained out of the hole.

Journalist, activist, and artist Abby Martin was seventeen at the time of the 9/11 attacks. An early interest in journalism was triggered by her high school boyfriend enlisting in the military after the buildings fell. Growing skepticism of the official justifications for the Iraq War influenced her journalistic writing and a ready embrace of "alternative" sources of information quickly led to her exposure to 9/11 conspiracy theories.

In a YouTube video, Martin gave an overview of her route down that rabbit hole at a time when she was deep inside it, at a March for 9/11 Truth on the streets of Santa Monica in 2008. The twenty-four-year-old Martin was the organizer of the San Diego 9/11 Truth Meetup group and an active member of the general 9/11 Truth community. While marching down Wilshire Boulevard carrying a sign saying 9/11 TRUTH Now, she was asked if and why she thought that 9/11 was an "inside job."

Absolutely it was! I know that because I've researched it for three years and everything that I uncovered solidifies my belief that it was an inside job and our government is complicit in what happened. . . . I saw the Pentagon, which confused me, and I started researching more, and I saw Building 7, and I saw the demolition of the two towers right in front of [my] eyes . . . it's someone telling you an apple is an orange, you just have to stretch your own mind to know what you are seeing is true.[3]

Martin was then asked what the most compelling evidence was of an "inside job":

Absolutely Building 7, without a shadow of a doubt. Building 7, no doubt in my mind. Building 7 was brought down by controlled demolition. Anyone who looks can see that. It's so weird, it's like black and white. Nobody can look at that building and say that was brought down by fire. There's nothing there, there's no debris. Even the Twin Towers, you look at ground zero, there's nothing there, there's just powder, where did a huge 110-story building go?

She then puts this in the context of her overall worldview, one where the people in power exert such a strong control over the media that basically everything is a lie, and people don't see what she sees.

It's just shocking, and the fact that they did it in broad daylight, it really is infuriating. And the people can't see it, because like the Nazi propagandists said, it's a lie so big that people can't see it, they can't believe it because it's so big, and so in your face.

I'm here, and I've been a 9/11 Truth activist for about three years, because I feel it's the most important issue to expose. They don't want any evidence to be uncovered about 9/11 Truth because that would ruin their whole system, they have everyone under this grid watching TV like zombies believing everything they see on television. So if they let one piece of evidence slip out then that could ruin their whole grid.

By 2014 Martin had largely backed away from the more extreme claims of controlled demolition and moved to a more reasonable position of simply criticizing the way the government responded to 9/11 and then exploited it. She appeared on Jesse Ventura's *Off the Grid* show at a time that Ventura was still pushing the more convoluted theories of controlled demolition, yet they did not discuss those theories directly.[4] In a rather bizarre juxtaposition, Abby Martin mentioned her prior belief in controlled demolition being brought up as a smear campaign, seemingly unaware that Ventura himself still claimed to hold those exact same beliefs, especially about Building 7.[5]

But what did Martin mean when she said, "I saw the Pentagon, which confused me, and [in 2005] I started researching more, and I saw Building 7"? It's very likely that she simply saw a combination of 9/11 Truth websites and 9/11 Truth videos like *Loose Change* (2005). Her comment about "someone telling you an apple is an orange" is a direct reference to a popular 2007 You-Tube video by Anthony Lawson that contrasts the collapse of Building 7 with conventional demolitions of similar-shaped buildings.[6] Her comments about the towers being "just powder" reflects many videos from that time (and to this day) that claim that the buildings were somehow turned to dust by unknown means like "nanothermite" or energy beams from space. Even her talk of the "big lie" is a reflection of similar claims in *Loose Change* and the 2007 *Zeitgeist* movie.

Abby Martin's route down the rabbit hole seems far more self-directed than Martin Beard's, but they both share that same key aspect of their journeys being both started and accelerated by the watching of seductive videos.

Your friend will almost certainly have had a similar initial experience at some point. They watched a video, that video led to another video, and they fell down the rabbit hole. That's the most common "how" part of the equation. The next thing we need to focus on is "why." Why did they get so easily sucked in? And if we figure out the how and the why, then what can we do about it?

Predispositions

One of the more common questions people ask about conspiracy theories and conspiracists is: "Why do people believe in conspiracies?" Is it a personality type? Something in the way their brains are wired? Some trigger event in their past? A reflection of their intelligence? Mental illness? The way they were brought up? Evolutionary brain functions? Something lacking in their education? The influence of their peer group?

While it's understandable to want to be able to say things like: "People believe in conspiracies because of X," the truth of the matter is that there's no simple answer for any one individual like your friend. Even if we consider wider populations there's no single clear correlation between any one factor (such as intelligence) and conspiracy thinking.

In a certain sense this question is not important. Something I've increasingly realized over the years, and something that has been greatly confirmed by the research I've done for this book, is that anyone can fall down the rabbit hole. It's true that certain factors seem to make it more or less likely to happen, but the key thing seems not so much the personality types or certain mental states of the individual, but rather their exposure to certain types of convincing information, like videos.

One simply cannot point to a single pre-existing factor, or even a constellation of factors, and say this is why a particular person fell for a particular conspiracy when they first began to be exposed to this new source of information. Conspiracy theorists are normal people just like you and me. But they are normal people who fell down the rabbit hole, whereas you and I were lucky enough to stay above ground.

That said, understanding how personality types and traits affect the speed and ease of one's descent into the rabbit hole is a useful part of understanding the entire picture of any individual. Perhaps your friend is actually one of those individuals whose conspiratorial thinking is closely tied to a particular mental quirk they have. Perhaps there actually is something that happened to them, or something about their current emotional situation that's a driving force in their obsessive researching and automatic distrust of authority.

Mental quirks, psychological abnormalities, are only a part of the picture. Everyone has mental quirks, but we also have normal aspects of the way we think that pretty much everyone shares. There's ordinary cognitive biases like the Dunning–Kruger effect, where people overestimate their own abilities. There's confirmation bias, where we tend to gather evidence that supports our theory and reject contrary evidence. There's biases from too much information, from selective memory, from pressure to make decisions, and from imperfectly filling in the gaps in information.[7] These are not aberrations, these are natural functions of the way our brains work. Biases are an inescapable part of human thought, and any honest thinker must give equal effort to both trying to avoid them in their own thinking, and to identifying them in the thinking of others.

An overview of current academic research on the topic can give us a very useful perspective. While psychological quirks might not be direct causal factors, they are certainly factors that could magnify the effect of seductive new information sources. They also help us understand why these new information sources (such as a website like Geoengineering Watch, movies like Oliver Stone's *JFK*, or videos like *September 11: The New Pearl Harbor*) work as well as they do. What buttons are they pushing? What types of person are they most appealing to? What ordinary biases do they exploit?

As well as providing us with useful perspective, an understanding of psychological factors also provides us with useful tools in talking to our conspiracy-minded friends. Try not to bring up the idea of specific psychological quirks with your friend, as that inevitably results in them thinking you are saying they are mentally ill. But you should certainly be able to bring up the idea of normal cognitive biases—precisely because they are biases that we all share.

There's an immediate type of common ground there that you can establish, particularly if you can give honest examples of your own experiences—for instance, many people will be familiar with personal examples of the Baader–Meinhof phenomenon (seeing or hearing something frequently after hearing about it for the first time), even if they have not heard the word before. You can share your own experiences ("I got a new car, now I see that car everywhere"), see if they will share theirs, and then maybe you can contrast that with how they never saw "Chemtrails" in the sky before they got interested in the Chemtrail conspiracy theory.

Academic research into the causes of conspiracism has been done for decades. In the US this academic interest was spiked by the conspiracist responses to key traumatic events in American history: the 1963 assassination of JFK, the 1993 Waco siege, the 1995 Oklahoma City bombing, and most significantly the 2001 9/11 attacks. The simple existence of these events is obviously one of the factors involved in why people believe, but they also provided a framework for academic studies into what led people to suspect there was some kind of government conspiracy behind them.

Academic Research into Conspiracy Thinking

One of the more commonly cited academic papers in recent years is the 2014 publication "Measuring Belief in Conspiracy Theories: The Generic Conspiracist Beliefs Scale," by Robert Brotherton, Christopher French, and Alan Pickering. The paper begins with a statement that you would be wise to keep in mind in the reading of any paper or popular article on the causes of conspiracy theories, or in listening to any supposed expert on the topic, myself included:

The psychology of conspiracy theory believers is not yet well understood.

A significant part of the reason why this is true is that psychology itself, the working of the human brain, is not yet well understood. We cannot understand the brain in the same way we understand a car engine, or even in the same way that we understand a complex computer program. Individual descriptions of events and feelings are often highly subjective. Many events and situations (like the way a person experienced 9/11, or their own family life) are unique and not repeatable. Everyone is different, and where we might make scientifically valid observations and deductions, these are generally statistical aggregates of many people. Sometimes these findings apply to the individual, but often they do not.

In the growing field of conspiracy theory psychological research in the 1990s and early 2000s there was another problem, a lack of a clear definition of a conspiracy theorist. How can we measure how a conspiracy theorist differs from the general population if we don't have a clear measure of what a conspiracy theorist is? Further, could research about low-level "Big Pharma" or "global warming" conspiracy theorists at one end of the spectrum even apply to "Chemtrails" and "Flat Earth" theorists at the other?

Brotherton attempted to bring some order here by creating a way of measuring of your position on a conspiracy spectrum, specifically the "Generic Conspiracist Beliefs (GCB) Scale," which gives a relatively simple measure of how much of a conspiracy theorist a person is. This is not the first or the only such scale, but it has quickly become a popular one used by many researchers in their studies.

Most of those studies try to determine if there is a statistically significant correlation between a pre-existing factor (e.g., social anxiety, intelligence, or personality type) and the subsequent development of conspiracism, as measured by something like the GCB scale. In many cases a small but significant correlation was found. Unfortunately, this type of study is often reported in the popular press as if that one factor was the *sole* causal factor behind the conspiracism. You'll get headlines like these (actual headlines):

- Studies Find the Need to Feel Unique Is Linked to Belief in Conspiracy Theories[8]
- Conspiracy Theorists Have a Fundamental Cognitive Problem [illusory pattern perception], Say Scientists[9]
- Losers Are More Likely to Believe in Conspiracy Theories, Study Finds[10]
- People Who Believe in Conspiracy Theories More Likely to Be Suffering from Stress, Study Finds[11]
- Narcissism and Low Self-Esteem Predict Conspiracy Beliefs[12]
- Conspiracy Theories: Why More Educated People Don't Believe Them[13]
- Study: The Personal Need to Eliminate Uncertainty Predicts Belief in Conspiracy Theories[14]
- Conspiracy Theories Mostly Believed by People on Far Left, Right of Political Spectrum[15]

It's quite straightforward! Your friend is a conspiracy theorist because they are narcissistic, suffer from stress, and have low self-esteem. They desire to feel unique and eliminate uncertainty; they tend to see patterns where there are none. They are poorly educated and are far left or far right wing, and recently were on the losing side of an election.

Clearly that's both overcomplicating things (it's unlikely all those factors are relevant to a significant degree in any one individual), and oversimplifying things. The oversimplification is the real danger here. If we were to take any one of the headlines alone (especially the versions in the tabloid press) it looks like scientists have discovered the sole reason people believe in conspiracy theories.

Even if you read the original work in depth, getting much out of these studies is still difficult. They are essentially taking a very complex situation and only looking at a very simple, narrow aspect of it. The subjects of the tests are often not particularly representative of the group of conspiracists that you might be interested in, and rarely will they match a particular individual like your friend. They represent a slice of society, sometimes in another country, and you can't extrapolate the results with much confidence. It just gives you a statistical correlation, often expressed in mathematic notion that is difficult to understand.

All that said, let's have a more detailed look at two of the more popular studies, and see what can usefully be applied.

The Need for Uniqueness

In "I Know Things They Don't Know!" French researchers Lantian, Muller, Nurra, and Douglas investigate if a motivational underpinning of conspiracy belief is simply the desire to be unique. Firstly, they surveyed people to find out if there was a simple correlation between belief in conspiracy theories and a sense of possessing unique knowledge. The participants (mostly French people in their twenties) were given an inventory test to rank them on the conspiracy belief spectrum (similar to the GCB scale, an inventory test just asks how much they agree with various conspiracy theory statements like "Lee Harvey Oswald did not act alone"). They then were asked to rank how unique the information was that they used to answer the questions, and if they got the information themselves or from someone else. Rather unsurprisingly, the higher on the conspiracy spectrum they were, the more unique they considered their knowledge, and the more they tended to think of it as knowledge they had acquired themselves. Lantian considered this as evidence that conspiracy theories would satisfy a need for uniqueness, but did not establish any causal link.

The second study set out to answer if people who had a chronic need for uniqueness ranked higher on the conspiracy spectrum. This second study was conducted on participants in the US using Amazon's Mechanical Turk system which allows a series of questions to be put to a large number of demographically specific people for a relatively low cost (thirty cents per person in this case). As before, the participants were ranked on a conspiracy spectrum (using Brotherton's GCB scale). Using a similar questionnaire technique, they are also given a measure on a scale of "Need for Uniqueness."

The results of this study were that: "A higher need for uniqueness was associated with higher belief in conspiracy theories." But how much? The short answer is "not a lot." The authors give a correlation coefficient "r" of 0.17. A correlation coefficient is a measure of how much two things can be statistically linked—the closer to 1.0, the stronger the link; and inversely coefficients closer to 0.0 reflect weaker links. This means the variation with need for uniqueness was associated with about 3 percent of the variation in belief in conspiracy theories.[16]

Not that impressive sounding, and a far cry from the headlines of popular articles reporting on this study. The popular science site IFLScience declared, "People Who Believe Conspiracy Theories Just Want to Be Unique, Say Psychologists."[17] It does not seem like they even read the paper though, as they were paraphrasing another site, PsyPost, whose headline read: "Studies Find the Need to Feel Unique Is Linked to Belief in Conspiracy Theories."[18]

Lantian's paper goes somewhat beyond establishing a simple correlation; in the next two studies they attempted to manipulate people's likelihood to believe in conspiracy theories. The third study was inconclusive, but the fourth appeared to show a possible causal link between conspiracy belief and a desire for uniqueness. This means if you encourage people to see uniqueness as a positive thing then they became more likely to believe in conspiracies than a person who was encouraged to think fitting in, or fighting for a common cause, was a good thing.

A causal relationship is important because it gets us closer to understanding what is actually going on. If we simply found that a need for uniqueness was correlated with belief in conspiracy theories, it does not necessarily follow that the need for uniqueness actually *caused* that belief, or even that it contributed towards it. The belief in conspiracies might have actually led to your friend experiencing the uniqueness of having special knowledge, and then enjoying that, and then seeking out more of the same. Correlation is not causation.

But the last part of the Lantian study does seem to suggest that in this case there is actually *some* causation. If they can manipulate people's conspiracy thinking via changing their need for uniqueness, then that means their need for uniqueness is, at least in some small way, leading to the conspiracy thinking.

But again, how much? It's hard to say because here the measure is still rather indirect. We are not measuring how having a need for uniqueness led to conspiracy thinking over several years. The measurement is of how much the conspiracy belief tendency of a person shifted immediately in the five minutes after being fed a few suggestive questions to make them think more highly of uniqueness. It's only measuring that one artificial moment in time.

But if we take this idea of the need for uniqueness from the academic into the more subjectively observed real world then there's something to this. While it's not entirely clear what people's need for uniqueness was before they got into conspiracy theories, it's very clear in a large number of cases of people who got in and then got out of the rabbit hole that a large motivation for them to *stay* down there, and a large motivation for them digging a deeper hole with more "research," was that they really enjoyed feeling special.

If this is indeed the case with your friend, then you will want to avoid anything that makes them feel ordinary. It is quite factually true that an inquiring mind is a great thing to have. That they are asking questions and not blindly accepting everything they are told is indeed something special. Try to convey this to them, but combine it with the need to also get things correct.

There's also the possibility that a need for uniqueness can be satisfied by becoming an advocate for science and reason. Many (but not all) of the stories in this book are from people who actually enjoy a different kind of special knowledge. They now know exactly why their old beliefs were wrong, and they know more about the world. As Willie put it:

I used to be entertained by conspiracy theories, but now I'm entertained by seeing them debunked.

They still get that little buzz from "I know things they don't know." But now they know things that are actually correct, which is even better.

"Losers"

One of the more unfortunate recent headlines regarding conspiracy theories has been "Conspiracy Theories Are for Losers." This is actually something of a play on words, the actual study being referenced is titled "The Effect of Conspiratorial Thinking and Motivated Reasoning on Belief in Election Fraud" by Edelson et al.[19] The study is not about what the word "losers" typically suggests (people who have failed at life though their own incompetence), but rather refers to the unsurprising finding that the people who identify with the political party that is not in power (i.e., that lost the most recent election) are more likely to believe in some conspiracy carried out by the party in power. Or more simply, conservatives are more likely to believe that Obama was a Kenyan, liberals are more likely to believe that Trump colluded with Russia.

More specifically Edelson's study was on voter fraud. Again unsurprisingly, the people who thought that the election was in some way rigged or unduly influenced tended to be the people who had just been on the losing side of the election—hence the unfortunate "Conspiracy Theories Are for Losers" headline.

The insulting phrase comes from one of the paper's authors, Joseph E. Uscinski, who coined it in a 2011 paper of the same name following a study of letters to the editor published in the *New York Times* between 1897 and 2010.[20] In that paper Uscinski and his co-authors say:

> We argue that perceived power asymmetries, indicated by international and domestic conflicts, influence when and why conspiracy theories resonate in the US. On this reasoning, conspiracy theories conform to a strategic logic that helps vulnerable groups manage threats. Further, we find that both sides of the domestic partisan divide partake in conspiracy theorizing equally, though in an alternating pattern, and foreign conspiracy theories crowd out domestic conspiracy theories during heightened foreign threat.

Which basically means that people are more prone to believe in conspiracy theories about things that they think threaten them, which includes both foreign threats and the election of politicians they voted against.

Uscinski continues on this theme in his 2014 book, *American Conspiracy Theories*, coauthored by Joseph M. Parent, arguing the most important factor in political polarization of conspiracy theories is the party of the president, and hence the popularity of any individual conspiracy theory will wax and wane with whoever is in power.

Anecdotally this comes through in 9/11 conspiracy theories. Since the events of 9/11 happened on the watch of President George W. Bush (a Republican) there was a tendency to blame it on a secret right-wing conspiracy. For the liberal friends of Willie this idea was so ingrained in their worldview that when he told them he no longer believed that 9/11 was an inside job, the first thing they asked him was, "Are you are Republican now?"

Former conspiracist Steve, who we will meet later in the book, also noticed something of a left-wing political bias in the early days of both the 9/11 movement and the early version of the Tea Party.

> *The initial part of the 9/11 movement was during the Bush administration. He'd got some characters there like Cheney who look kind of sinister, so that made it more plausible. Then the changeover from the Bush administration to the Obama administration lessened it to some degree because a lot of the group were more liberal.*
>
> *The first Tea Party was in Santa Monica, California, and it had a lot of liberals in it. There were lots of anti-banking types, and 9/11 Truthers, only half of them were that conservative, it was a weird mix. The initial thing was how the people behind 9/11 were making environmental laws where the government could take over your house, or your entire life. The Tea Party is a conservative thing now, and that's also how many of the conspiracy people are going.*

Is this political polarization of conspiracy theories a useful observation? Firstly, you should avoid using the phrase "conspiracy theories are for losers," even in jest and even if you immediately explain it. Conspiracy theorists feel bad enough about being called conspiracy theorists, so if you wave the label "loser" anywhere near them then it's going to be a problem, causing them to raise mental barriers against whatever you say next.

If you want to bring it up with them, then you should explain the entire thing in one sentence: "People who supported the losing side in an election are more likely to believe conspiracy theories that are supposedly orchestrated by the winning side."

Ultimately though, there's not much to be gained in discussing the "loser" theory with the conspiracy theorists themselves until they are already on their way out of the rabbit hole. When they are deep in, this idea is going to seem to them like an attempt to belittle their concerns. Unless you yourself have been through some similar phase in the past (from which you recovered), then it's not going to help.

On the other hand, if your friend is emerging from the rabbit hole, or especially if they are mostly out, then a degree of healthy introspection can help reinforce their new position. If their conspiracy theories were indeed largely focused on an opposing political party (like 9/11 and Bush, or gun control False Flags and Obama), then understanding that the seductiveness of the ideas was magnified by being on the losing side will go at least some way towards reinforcing their realization that the seductiveness was not based purely on facts.

Don't Pigeonhole

Ultimately, this interesting academic research is just that: interesting, but largely academic. It may well be true that people are slightly more likely to believe in conspiracy theories if they have a need for uniqueness. But, other than pandering slightly to that need, it's not particularly useful information.

Other studies are even less useful on the individual level. Now to be fair, the purpose of these studies is generally to look at things in the aggregate, and practical benefits may be derived from them if we are to apply them to larger scale things like education policy, or even writing books on debunking. But when dealing with the individual it's quite hard to know if they even fit in a particular pigeonhole, let alone do anything about it.

For example, in the paper "Fake News: Incorrect, but Hard to Correct," De Keersmaecker and Roets note three findings:

1) *When people learn their attitudes are based on false information, they adjust them.*
2) *People low (versus high) in cognitive ability adjust attitudes to lesser extent.*
3) *Adjusted attitudes remained biased for people low in cognitive ability.*[21]

This is saying: "You can change minds by debunking, but it's harder to do with stupid people." Now the first part is heartening—at least we know that presenting missing information has an effect—but it's of no real use to be told it's harder to do with people with low cognitive ability. If our friend has low cognitive ability, then what should we do, give up? (You could try to increase their cognitive ability, but that's rarely going to be a practical approach.)

Similarly, papers like "Does Self-Love or Self-Hate Predict Conspiracy Beliefs? Narcissism, Self-Esteem, and the Endorsement of Conspiracy Theories" are rarely helpful at the individual level.[22] The degree of correlation is generally small, and they deal with factors that are difficult to gauge in your friend. Note here that "narcissism" and "low self-esteem" are opposing factors. What if we think our friend is a bit narcissistic, so we push things in the opposite direction, only to find we go over into low self-esteem. It's probably going to be a waste of time.

Let's leave the academic aside, avoid pigeonholes, and look at the practical. How do real people actually emerge from the rabbit hole?

Routes Out of the Rabbit Hole

In discussions with former Truthers there seems to be two fairly distinct ways people find their way out of the rabbit hole: either a gradual change of perspective or a sudden realization of the fundamental flaws in their position. Sometimes it is a combination of the two.

Escapes by changes in perspective often start with a seed of doubt—they discover that some foundational piece of their conspiracy theory was incorrect, and they begin to look at other parts of the theory that they had assumed to be true. But it can also simply come from learning more about how the world works and gaining more life experience. They emerge a bit at a time, slowly learning new things until they find themselves on the other side of what they thought was an impassable line.

Escapes by sudden realization come only after a build-up of new information that is initially strongly resisted. While they are learning new things, they are rejecting those things as false disinformation. Eventually their knowledge of the evidence against their theory builds up and leads to a more sudden realization they were wrong, a breaking of the dam, and a rapid movement over their own demarcation line.

But there's a single prime mover here in both routes: *exposure to new information*. Conspiracy theorists flourish in walled gardens. When asked where they get their news they will often point only to alternative fringe sites like Alex Jones' Infowars, or more esoteric conspiracy theory sites like Rense .com, or even David Icke's reptilian Illuminati related news.

Cass Sunstein and Adrian Vermeule describe this as a "crippled epistemology"—simply defined as having "a sharply limited number of relevant informational sources."[23] You can be a new information source for your friend, but it's equally important that you introduce them to (or help rehabilitate) other relevant information sources.

These two main ways in which exposure to new information (and information sources) can help people out of the rabbit hole can be seen in two escaped 9/11 Truth activists: Abby Martin and Charlie Veitch.

Abby Martin seemingly changed her mind about 9/11 in the more gradual way, by gaining perspective. In 2008, she was a supporter of the 9/11 Truth movement and described the 9/11 attacks as an "inside job" as she participated in a 9/11 Truth march in Santa Monica, California. In 2012 she moved to Washington, DC, to work for RT America. This gave her access to a lot more people in varying positions of power and allowed her to observe "how the government really works." Her discovery of Washington as a corrupt yet lumbering bureaucracy simply did not fit with the type of super-competent, all-powerful evil entity required to pull off the version of a 9/11 conspiracy theory she previously subscribed to. It seems she emerged slowly, with a gradual realization that her previous belief made no real sense in the context of the world as she now understood it.

Veitch was an energetic and outspoken Truther for several years, seemingly deep down a 9/11 rabbit hole he got into via the Alex Jones movie *Terrorstorm*. His exposure to new information came suddenly over a few days in a road trip organized by the BBC. He met many of the people intimately involved in 9/11—from the architects of the Twin Towers, to the first responders, and the families of the victims. They told him face to face what really happened to them, what they knew, and what they thought happened. For Veitch, this was simply information that he had no access to before, although he was aware of the general objections to his theory. The things they were telling him made sense to him, and after some initial emphatic resistance he quickly recognized the wrongness of his previous position. He emerged from that rabbit hole rapidly, and nearly completely, over the course of a few days.

The routes out vary, but the bottom line is that exposure to new information and new perspectives does change minds.

Where People End Up

Just as the routes in and out of the rabbit hole are varied, there's a variety of different places people can end up after they finally get out. For some people the change is a binary one, a near 180-degree flip-flop from believer to skeptic. Willie is now entertained by debunking where he was once entertained by conspiracies. Later we will meet Steve who was once a hard-core conspiracy theorist, a believer in 9/11 controlled demolitions. But he changed, fairly rapidly, and went from being an attendee at conspiracy marches to someone who went to them simply to try to convince the attendees (many of whom knew him) that they were barking up the wrong tree. Like some former smokers, or former religious people, his reaction to discovering a new truth was to become as passionate and evangelical for that truth as he was for his old "truth."

Not everyone makes such a polarized switch; some people keep at least one foot in the rabbit hole. A 9/11 conspiracy theorist might go from thinking for sure that a missile (not a plane) had hit the Pentagon, to thinking that there was simply a lot wrong with the official story. Sure, maybe the towers were not destroyed by pre-planted explosives, but they still have questions about who the hijackers really were, who really planned 9/11, and who paid for it. The certainty is gone, but the suspicion, and sometime the investigative fervor, remains. It's just a shift in where they draw their demarcation line.

Many people who emerge from the rabbit hole simply lose interest. They largely drop the subject of conspiracy theories from their lives and move on. Like Scott, a participant on the BBC's UFO episode on *Conspiracy Road Trip*, after being intensively deluged with new information for several days:

I came here thinking that I might get some kind of facts or answer, but I haven't got anything, which makes me think that there isn't an answer. I don't want to sit around anymore thinking what's true and what's not. You know, I'd rather just forget about it, and focus on other things.[24]

Recognizing that his own theories regarding the cover-up of UFOs did not really hold water, Scott also seems to recognize that reality is complicated and not always amenable to straightforward explanations. This is, in a way, a very honest and self-aware realization. We can't explain everything, we can't understand the math and science behind everything that happens any more than we can know what goes on behind closed doors in the upper levels of government. While it's great to be more interested in science, it's also a perfectly valid position, shared by many scientists on different topics, to simply say that you don't know the answers to everything. Some people continue with an interest in politics and corruption, but it's not for everyone. Do what you can, but recognize your limitations, and get on with your life.

Regardless of how people get in and out of the rabbit hole and where they end up, the bottom line is that their escape is for them a highly positive and often life-changing experience. I've had many emails, some from the people whose journeys are chronicled in this book, thanking me for helping them out of their conspiratorial mindsets.

Let's move on the next stage: how exactly *do* we help people escape the rabbit hole? What are the best practices and practical steps?

CHAPTER FIVE

Core Debunking Techniques

There is no magic bullet. The conspiracy theory rabbit hole is a complicated place, and people are complicated individuals. The reasons why people get sucked into the rabbit hole differ. Once in, their rate and route of descent into the deeper levels of the hole can vary immensely. Similarly, the paths out of the rabbit hole vary by individual. When shown something they believed to be true was actually false, some people simply accept they were wrong and move on. Other people respond in entirely the opposite manner, taking the claimed falsification of evidence as more evidence that they were right all along. Some people get out quickly, some pop out suddenly after years of resistance, and some emerge gradually, almost imperceptibly.

There's no simple step by step technique you can use that is guaranteed to get your friend out of the rabbit hole. But there are things that have worked for many people, which we can boil down to three areas:

1. Maintain an effective dialogue
2. Supply useful information
3. Give it time

You could quite accurately paraphrase that as, "Talk to them, show them stuff they missed, and don't rush." These three areas can be broken down further.

Aspects of maintaining an effective dialogue
- Understand what they are thinking and why
- Be respectful, honest, open, and polite

- Find common ground
- Validate their genuine concerns
- Avoid the backfire effect

Aspects of suppling useful information
- Show them mistakes they have made
- Show them mistakes their sources have made
- Show them things they have missed about the topic
- Show them other information that helps them gain perspective

The final point, "Give it time," is as simple as it sounds. These things take time. People very rarely just flip a deeply held belief. You might have quite an extensive discussion with a believer on all aspects of the 9/11 controlled demolition theory and it may seem like you are getting nowhere, or even going backwards. You might come back to them in a few days and still find no change at all. You may repeat this for weeks. Not everyone is reachable, but those people who *have* escaped from the rabbit hole almost always describe it as a lengthy process—something in which the role of a friend or other source of useful information was important for some time before they really started their journey out, and was important for some time during and after that journey. Give it time.

This simple method of effective communicating and supplying useful information is based on what I've learned writing at Metabunk over the last eight years (and in various other sites for a decade before that). But it's not something novel or unique. Many aspects of the method have come from other sources and are simply time-tested techniques of effective communication with people holding firm beliefs.

This is not a step-by-step recipe; you don't have to do one thing before the other. A lot will depend upon your friend, where they are in their beliefs, and how they respond to criticism. For some people the direct contradiction of their beliefs can trigger fierce rejection, and it's better to more subtly work on establishing common ground, widening their perspective, and supplying more neutral information.

There are a lot of different conspiracy theories. There are a lot of variations of those theories. There are a lot of different people who believe those variants in different ways. Your friend is unique, and general rules of thumb might not apply, or might even backfire. The Metabunk method is a very loosely structured toolbox of techniques, guidelines, and information where you have to pick the best tool for the job and adjust that approach to the unique circumstances of your friend's particular rabbit hole.

Develop Understanding

You can't show someone where they have gone wrong if you don't actually understand what it is they believe. A foundational part of effective debunking is coming to grips with what you are debunking. A lot of this book is simply about gaining that understanding in order to facilitate good communication. It's far too easy to dismiss people as silly conspiracy theorists and assume that the solution is simply going to be to show that person the real facts that explain their theory.

In reality it's a lot more complicated. I've been talking to conspiracy theorists for many years, and I'm still encountering variants on the various theories. In part this is because the theories have always had a broad spectrum of different levels and versions. But it's also because the theories keep evolving. For example, the various 9/11 controlled demolition theories went through a variety of stages over the years: there was the original simplistic "free fall" and "laws of physics" arguments (based largely on the collapses looking weird). Then a variety of arguments about odd things like the amount of dust or the shape of the dust clouds emerged. Then there were things like pools of molten metal and analyzing the dust for evidence of nanothermite. Over the last few years, there's been a focus on the minutiae of the Building 7 collapse. Was a particular girder-supporting bracket eleven or twelve inches wide? Was that girder bonded with the concrete? How hot exactly did a certain beam get, and when did it get that hot?

The first step in gaining an understanding of what exactly your friend believes is to listen to them. The next step is to ask them non-judgmental questions. This generally goes one of two ways. Either they will refer you to something they consider to be an authoritative source (like Architects and

Engineers for 9/11 Truth, or a video like *September 11: The New Pearl Harbor*), or they will start to list the various pieces of evidence that they feel are most compelling.

Either way, listen to what they are telling you. If they recommend a video or website, then ask them what was in it that they found the most compelling evidence. Ask them what was the *first* thing that got them interested in this theory. What's a top five list they might give to someone? How about a top two? Watch the video if you have the time. Read some of the website. It's a good investment in trust and will give you an overview of their beliefs.

To effectively communicate you also need to know where they are on the conspiracy spectrum. Don't just look at what they personally believe. Examine the various versions of the theory. You want to find their demarcation line—what is the more extreme version of the theory beyond which they think it's just silly? What claims are only just on this side of their line? What's just on the other side?

As well as figuring out the specific variation of their favorite theory, you also want to figure out what their understanding of the broader world is. What are the gaps in that understanding? What areas do they know more about than you do? Try to understand how the theories like 9/11 or Chemtrails actually fit into that broader image they have of how the world works. Who do they think is behind the conspiracy? Who benefits? How does this benefit work? Ask them this not in a critical way, but in a genuine attempt to get a better understanding of their worldview.

Understanding all this is vital to both stages of the Metabunk method: effective communication and supplying useful information. To be effective in communicating you need to know their mental map, the context in which your communication will be received. To supply useful information you need to know what the gaps are in their knowledge of the world and in their conspiracy theories. You also need to figure out what it will take for them to even consider information that could fundamentally negate their worldview.

Foster Trust and Respect

Understanding your friend's beliefs about their conspiracy theories is key to communication, but equally important is trying to understand your friend's beliefs about *you*. Rather surprisingly, their view of you might be very similar to your initial view of them. They probably consider you to be someone who has been fooled by bad information into believing an impossible story (like cave-dwelling Arabs with box cutters defeating the US, the government *not* poisoning us, or the Earth being round). This mirroring of positions might go so far that they actually consider *themselves* to be the reasonable debunker, and they are actually trying to explain things to you, the confused friend.

It's important not to push back against this position. Respect it. If they want to try to convert you then that's a great way of getting the conversation going. You can make it clear from the start that you don't really believe their theory, but you can (honestly) say that if there was some compelling evidence, then you would certainly consider it. Give them the opportunity to convert you. This opens the door for them to explain why they believe, and if you genuinely listen to what they say you will gain a very useful perspective, and also increase the odds that they will later also genuinely listen to you. If you respect them, and make an effort to understand their argument, then they will appreciate this, and in turn will respect you more. They will probably have had many situations where their ideas were flatly rejected or laughed at, and so being treated with respect will go a long way towards gaining their trust.

An open-minded two-way discussion is the best-case scenario. Unfortunately, it's quite possible that instead of them considering you simply uneducated in the facts of the matter they might consider you to be a shill, someone who is pushing a fake official story for personal gain. I've been debunking as a hobby for a very long time, so I've gotten this response quite often. As mentioned, my approach has always been to be as straightforward and consistent as possible. I stay polite and respectful, and I work on developing a good relationship with them based on an honest exchange of information. The longer you interact with them like this, the more likely they are to see that you are not a shill. Consistent honesty is the best way to establish trust and respect.

Finding Common Ground

Any argument an individual makes is built upon a huge and complex web of personal knowledge and beliefs, both true and false. You have one huge and complex web, and your friend has a different one. You cannot have a useful discussion with your friend if there's no area in which your web of knowledge overlaps their web of knowledge.

Perhaps the most common mistake that people make when trying to debunk something in a person-to-person setting is assuming there *is* common ground where there is not. You are automatically forming a set of mental assumptions about what your friend thinks. This is a mixture of both over-estimating their knowledge (for example, assuming your friend knows what a chemical compound is), and underestimating it (for example, assuming they do not know that homeopathic remedies are heavily diluted). This leads to ineffective communication, and in the case of underestimating it could lead to a feeling of being insulted and belittled that will cause your friend to grow defensive and suspicious.

It's vitally important to establish true common ground and a shared understanding of where the differences actually begin. In his book *Intuition Pumps and Other Tools for Thinking*, Daniel Dennett gives a three-step process designed to narrow in on this ideal state.

- *Re-express their position better than they do themselves*
- *List points of agreement, especially uncommon points*
- *Mention anything that you have learned from them*

Step one is to re-express your friend's position better than they did themselves. This will do two things. Firstly, it will show your friend that you understand what they are trying to say and are taking a genuine interest in it. Secondly, it will show them that you are not trying to belittle their argument by making it seem silly. By presenting an even better version of their argument it raises the bar for you, and if you go on to refute their argument then the refutation will have much more validity—both because you've shown you understood it, and because you actually improved it before you analyzed it.

This discussion technique is known as the "principle of charity." On Wikipedia (a forum beset with differences of opinion) a related concept is "assume good faith." By giving your friend's argument the best possible interpretation, you strip away your own bias to simply reject it or even laugh at it. You do your best to try to make their argument and their evidence actually work. This guarantees that when the flaws in their argument are revealed, they will be much closer to the bone, and based far more on a shared framework, and hence it will be far more likely your friend will take them seriously.

Step two is to list the points of agreement. This can be a gradual exploratory process, as it is not always going to be immediately apparent what the points of agreement are. You might need to do a little hunting to find out where the boundaries of disagreement are. You might want to start with something that's very uncontroversial but is directly related to the topic at hand. For example, with Chemtrails, you could ask them if they are concerned about pollution around airports. If they are, then you can share your own level of concern (if any). I used to live ten miles from Los Angeles International Airport, and just a mile from Santa Monica Airport. The noise from both bothered me, and I did have some concern about exhaust emissions contributing to air pollution.

If the topic of discussion is highly contentious, then it might be better to discuss an unrelated topic first, just to feel out the general parameters of the discussion, and to allow you both to get to know each other's mind a little. For example, when I met with a well-known Chemtrail promoter, we first talked about the pharmaceutical industry, and how we both felt there were areas where it was putting profits too far above the well-being of people— particularly when it came to marketing of drugs for what seemed almost like invented conditions, like chronic social anxiety. Talking about this topic, and not agreeing on everything, helped break the ice, and to show both of us that the other was willing listen, and to disagree without malice.

Step three is to mention anything you might have learned from your friend. This serves multiple purposes. It increases rapport, as your friend will feel they are actually communicating with you. It also creates in your friend's mind a better overall picture of what you know and understand. Fundamentally, though, it will cement the common ground. You learning things from them lays the foundation for them learning things from you.

Validate Their Genuine Concerns

It may seem counterintuitive to suggest that conspiracy theorists have genuine concerns. Indeed, the ridiculous nature of many of the more extreme conspiracy theories makes it tempting to dismiss the theorist off the bat. But when we talk about the "genuine concerns" of your friend, we do not mean the extreme claims in their theory. Instead we are referring to real-world concerns that are related to the domain of the theory.

Consider "Chemtrails," a seemingly ridiculous theory about a secret decades-old plot to alter the climate by spraying things out of planes without anyone noticing or complaining. There are actually genuine concerns in this area. Planes (like most vehicles) are sources of pollution. The contrails that planes leave behind are just composed of water, but they are a stark reminder that the plane's exhaust is spewing things out into the upper atmosphere. Planes contribute to atmospheric pollution in the form of small amounts of toxic compounds, and large amounts of carbon dioxide. Contrail clouds can turn an otherwise blue sky milky white, a form of *visual* pollution. They can even contribute somewhat to climate change by (inadvertently) trapping heat at night.

The topic of geoengineering itself contains many genuine concerns. There's the uncertainty of the potential side effects on the climate, on human health, and what will happen if we do it and then stop. Rose Cairns, a science policy researcher, notes that the Chemtrail believers who think that Chemtrails are a covert way of implementing geoengineering actually are quite right to be concerned about geoengineering, even if they mistakenly think that it has already started.

> Similar logics, concerns and fears animate both the Chemtrail discourse and wider discourses of fear about the climate . . . [there are] a number of ways in which the Chemtrail narrative may contain important insights and implications for the emerging politics of geoengineering that cannot be dismissed out of hand as 'paranoid' or 'pathological.'[1]

If a Truther tells you that 9/11 was used as a pretext for war with Iraq, or a Chemtrailer tells you they think geoengineering is too risky to rush into, or a Flat Earther tells you that we put too much blind faith in the authority of science, then tell them that yes, you can see their point, and you might even be able to find some areas of agreement.

This validation need not even be something you personally agree is a problem. Perhaps they dislike how contrails sometimes cover the sky. This might not bother you, but you can still see how someone might be bothered by it. It might be something that you've never even thought about. Either way, you can tell them at least that you see their point. While it is not common ground in the sense of both agreeing with the same concern, it's still common ground if you mutually get to understand the point of view of the other.

On a broader level, something underlying most conspiracy theories is a profound distrust of people in power. It's very important to explain to them just how much you yourself share in that distrust. Be perfectly honest here, tell them who you do or do not trust, how much you trust them, and why. Distrust, or at least suspicion, of authority is a healthy quality to have.

To the extent that you share their concerns, then validate those concerns. Where you don't share the same degree of concern, then try to understand why they are so concerned and explain your understanding to them. If you fundamentally disagree with their distrust, then at least acknowledge and discuss that distrust and your disagreement, and then use that shared understanding of the situation to build a constructive exploration of the actual evidence.

Illuminating the Rabbit Hole

For people to escape from the rabbit hole they need to be able to see where they are, where they might get to, and how to get from one place to another. Carl Sagan famously described science as "A Candle in the Darkness," in the subtitle of his book *The Demon-Haunted World*.

> In every country, we should be teaching our children the scientific method and the reasons for a Bill of Rights. With it comes a certain decency, humility and community spirit. In the demon-haunted world that we inhabit by virtue of being human, this may be all that stands between us and the enveloping darkness.[2]

Candles are great, but we need more than candles. We need flashlights, we need spotlights, we need floodlights. We need emergency exit row lighting, night vision goggles, flares, flash-bangs, and flaming torches. There's a lot of people down a lot of rabbit holes, and lighting a few candles isn't going to cut it.

We've seen how some people, like Willie, can identify a pivotal point, a key piece of information they were shown that led them to start questioning other things that they thought were true. Other people, such as Abby Martin, had a more gradual emergence as they slowly learned more about the real world. I refer to these two variations of illuminating the rabbit hole as the *spotlight* and the *floodlight*.

Spotlight Debunking

The spotlight technique is to shine the light on one single claim of evidence, or one particular claim. There are certain core beliefs that most people in a particular conspiracy share. In the case of Chemtrails, the most common core belief is that contrails cannot persist more than a few minutes. This belief is false, and in this case we have a very powerful spotlight in the form of many old books on clouds, each of which mentions contrails and each of which says that contrails can persist for a long time if conditions are right.

I have a personal collection of about twenty different such books. Using them I posted a four-minute video to YouTube back in 2014. I very simply explained what people were saying about contrails being short and Chemtrails being long, then I just went through seventy years' worth of books on clouds, said what date each was from, and read the section on how long contrails should last.

This shone a bright spotlight on that one core claim. It's possible for the believer to double down and say that Chemtrails have been sprayed for decades, but most people who think that contrails can't persist will combine this belief with their own personal memories of not really remembering seeing contrails before hearing about the Chemtrail theory. But at the very least the spotlight has revealed a problem with their theory. If trails can persist then why do the Chemtrail promoters say they cannot? And if they cannot persist then why do seventy years of books say they do?

Carey Dunne of the UK newspaper *The Guardian* used this spotlight with a Chemtrail-believing couple in Northern California.

> *I play Tammi and Rob a YouTube video by Mick West, who runs the conspiracy theory-debunking blog Metabunk. Going through 70 years of books on the science of clouds, West explains why, depending on atmospheric conditions, contrails can either evaporate rapidly or persist and grow into sheets of cirrostratus [clouds].*
>
> *After this show-and-tell session, Rob claims "nothing will change [his] mind," but Tammi says the video in particular put her "on the fence."*[3]

Being immediately put on the fence by the spotlight might seem very encouraging, but the real test is how the effect plays out over time. Dunne returned two months later. True to his word, Rob was still unswayed, but Tammi still seemed somewhat open to the possibility, at least intellectually.

> *Tammi, though, says the facts got her "questioning. If I wasn't so busy farming, I'd do more research," she says. "I need more information. But then when I see it, heavy in the sky, I think, there's no fucking way that's not Chemtrails. I never saw clouds like that as a kid. My gut and heart still tell me something's going on."*

The partial failure there was due to two factors. Firstly, the spotlight was not held there long enough for the explanation to really sink in. But more significantly here is the fact that it is difficult for a couple to change their beliefs. Partners want to be supportive of each other, so there's a natural reluctance to let your own beliefs change in a way that would contradict the beliefs of your significant other. The spotlight can be shaded by someone with their back to it, protecting their partner.

There are several other core claims in the Chemtrail theory. They might not always be the ones you expect. The ballast barrel claim (where ballast barrels on test planes are misidentified as "Chemtrail" barrels) seemed like a simple little thing to me, trivially easy to explain. I wrote and updated a long post on Metabunk with many examples of mislabeled aircraft barrel images. I did not really expect it to have much effect—it was just a fun and easy thing to research.

Hence, I was quite surprised when Willie told me that not only had the ballast barrel claim been key to his going down the Chemtrail rabbit hole, it was also key to him getting out again ("it changed my life"). Perhaps it was the sheer simplicity of the claim that did it. If he had been tricked for *years* by something that was easy to debunk, then there surely must have been other things he'd been tricked about, and other things that he'd been lied to about. As soon as he started looking for these things with a truly open mind they started to become apparent.

The spotlight need not shine incredibly bright, it just needs to shine in the right place for that person. You might not get many opportunities, so if you are going to take the time to shine a light for someone, then spend a bit of time in figuring out what will work best for their particular beliefs.

Floodlight Debunking

Some people can begin their journey in and out of the rabbit hole by pivoting on one key piece of information, like the long contrails, or barrels on planes. Others though are more convinced by the sheer *weight* of the evidence. The seemingly vast amount of supporting evidence for a conspiracy theory (often referred to as "proofs"), and the corresponding lack of opposing evidence, is what keeps them in. The people who have this broad base of belief have what Sunstein and Vermeule called a "crippled epistemology," a narrow and restricted set of sources of information.

For many conspiracy theorists being shown that one particular "proof" is wrong is irrelevant. For a start they often don't really believe your explanation, but even if they did, it's irrelevant to their belief because each and every "proof" is to them a near absolute actual proof *by itself*—meaning they only need one thing, one proof, to continue to believe.

Floodlight debunking means we shine a light on everything. If they've got a list of 200 proofs of a Flat Earth, then it might actually work best to respond to all 200 items—or at least find someone else who has. If it's 9/11 then you might not actually get anywhere until you've refuted a substantial chunk of the *Loose Change* documentary, or you've addressed all ten of Architects and Engineers for 9/11 Truth's (AE911Truth's) "10 Key Points" about the towers, and the "25 Points of Specific Concern in the NIST WTC Reports."

Floodlight debunking also means bringing perspective. Conspiracy theorists often have a warped view of the way the world works and how power and government actually work. So try to examine the everyday forms of political corruption that actually go on, such as campaign finance and lobbying. With 9/11, look at actual conspiracies, actual real-world "false flag" events, and suggested events like Operation Northwoods. Compare those to what your friend suggests happened to the Twin Towers. With Chemtrails look at the actual state of climate engineering research and ask them to view their theory in that context. Would we be geoengineering if we don't even know what it does? With Flat Earth you can get some actual perspective by going to the beach and looking at distant islands obscured by the curve of the Earth.

The most valuable perspective is simply to understand how society functions. How is wealth created? How are laws created? How do elections work? Who's actually on the *Forbes* list of billionaires? How is science funded? How many scientists and science students are there? Why are there wars? How did the Russian oligarchy arise? What do other countries think about the US? What do people talk about at Davos?

You may well be hampered here because your friend will initially feel they understand the true workings of the world far better than you do. This might actually be true in some areas, but less so in others. Try to approach it as a mutual journey of discovery. You will probably learn something from them. Embrace this, and they will be more open to learning from you. Both of your perspectives will be expanded, and the exit from the rabbit hole will get a little closer.

Be Polite and Respectful

If your goals are to effectively communicate, then you need to be polite and you need to respect your friend.

People will push back when they feel they are being attacked. *Regardless of the intentions* of the person they are talking to, or the article they are reading, or the video they are watching, if they feel it denigrates them in any way then they are far less likely to actually consider the validity what is being said.

Phil was a visitor to my Chemtrail debunking website ContrailScience .com who did not agree with this approach. Phil was very intelligent, he knew a lot about the science behind contrails, and he was aware of most of the problems with the Chemtrail theory. He was able to provide concise explanations that refuted the bunk posted by others and gave helpful overviews of the actual science.

Yet Phil would almost invariably conclude his posts with something along the lines of "get an education," "take some classes," or "that's just ignorant." He would sometimes use more direct insults like "uneducated chemtards," "uneducated simpleton," and even "mentally ill," and urged people to "get professional help" for their "paranoia."

This had two effects. Firstly, it meant either that any Chemtrail believer who engaged Phil in conversation would leave the forum or that the conversation would almost immediately degenerate into a flaming insult session where the original topic was essentially forgotten. It was highly counterproductive.

Secondly, it tainted the site, and it tainted the other debunkers by association. Since Phil was quite an active poster, it was very easy for a visitor to get the impression that his opinion was the opinion of the site. The opinion came across as Phil being an intellectual snob, arrogant and disdainful of contrary opinions. It also came across as Phil being totally unwilling to listen to the opinions of the believer. Since the believer was strongly emotionally and intellectually invested in their beliefs, this dismissal came across as a slap in the face. A direct insult. When more polite debunkers tried to explain things, the damage had already been done. The shields were up, and the mind was closed.

I tried to explain this to Phil several times, but he in turn took my criticism as a direct insult. So I banned him. This was not an action I took lightly, as he was generally a good contributor. But the damage he was doing by his insulting manner was outweighing the useful factual contributions that were great alone, but unfortunately seemed only to be a preamble to his insults.

Don't be like Phil. Even if you feel your friend is being stupid, uneducated, or even crazy, it's still best to just focus on the facts. *Show* them where they were wrong, show them what they missed, show them where their sources are wrong. Don't tell them they are stupid. Be polite, please!

Avoid Backfire

The backfire effect is an occasional phenomenon where attempting to correct a false belief actually "backfires" and makes it stronger because the person involved fights against the correction. Anyone who has attempted to debunk things online will have had the experience of presenting very reasonable counterevidence to someone, only for them to seem to become even more entrenched in their position.

Most research into this effect is relatively new and has revolved around politically charged beliefs such as climate change. In 2010, Nyhan and Reifler wrote:

> Can these false or unsubstantiated beliefs about politics be corrected?...
> Results indicate that corrections frequently fail to reduce misperceptions among
> the targeted ideological group. We also document several instances of a "back-
> fire effect" in which corrections actually increase misperceptions among the
> group in question.[4]

Shortly after that Lewandowsky and Cook wrote *The Debunking Handbook*, discussing ways to avoid this effect, focusing on the issue of climate change denial. They cover three strategies:

> Debunking myths is problematic. Unless great care is taken, any effort to debunk
> misinformation can inadvertently reinforce the very myths one seeks to correct.
> To avoid these "backfire effects," an effective debunking requires three major
> elements. First, the refutation must focus on core facts rather than the myth
> to avoid the misinformation becoming more familiar. Second, any mention of
> a myth should be preceded by explicit warnings to notify the reader that
> the upcoming information is false. Finally, the refutation should include an
> alternative explanation that accounts for important qualities in the original
> misinformation.[5]

The idea behind this advice is that to debunk a topic effectively, you've got to first, avoid mentioning it; second, take pains to preface it as being wrong before you do mention it; and third, provide something else to fill the mental gap you create when you debunk the misinformation. (They later add a fourth recommendation, which is that information should be conveyed graphically if possible—a good idea.)

This advice on avoiding the backfire effect is *basically* sound, but it's important not to jump through hoops unnecessarily. One should not avoid discussing the misinformation to the extent that your friend does not know what you are talking about. Focusing only on facts is great, but it's also important to understand why they believe as they do, which means you may also have to discuss the false belief. Providing an alternative explanation to fill the gap is often very useful, but if you find yourself struggling to create one, then perhaps "that's incorrect because . . ." is actually the best approach.

For example, if your friend thinks that HAARP (High Frequency Active Auroral Research Program, a research facility in Alaska) causes earthquakes, then you could avoid any backfire effect by solely explaining what causes earthquakes—plate tectonics. But to fully address the issue you also need to explain why HAARP is incapable of creating earthquakes—it's a low-power radio transmitter that can only affect one spot of the sky above Alaska.

You should exercise caution in applying lessons derived from surveys to the individual situation of your friend. Is the backfire effect even a problem for them specifically? In 2018, Wood and Porter published "The Elusive Backfire Effect: Mass Attitudes' Steadfast Factual Adherence," in which they attempted to extend Nyhan and Reifler's original research and were somewhat surprised to find that a generic backfire effect did not seem to exist.

> *Across all experiments, we found no corrections capable of triggering backfire, despite testing precisely the kinds of polarized issues where backfire should be expected. Evidence of factual backfire is far more tenuous than prior research suggests. By and large, citizens heed factual information, even when such information challenges their ideological commitments.*[6]

On the other hand, both the backfire effect and the heeding of factual information were somewhat confirmed by Chan, et al. in a 2017 meta-analysis, where they found:

> A detailed debunking message correlated positively with the debunking effect. Surprisingly, however, a detailed debunking message also correlated positively with the misinformation-persistence effect.[7]

The takeaway from Chan is not so much that you need to provide an alternative explanation, but you need to provide detailed factual reasons why their explanation is wrong. You need to exercise caution, because people still respond irrationally when motivated to do so and people do push back against what they see as attacks on their ideology, but people simultaneously *do* heed factual information. Chan describes the approach as "detailed debunking," which tallies with the second broad part of the Metabunk method: supplying them with useful information.

In my experience, backfire is largely the result of problems with communication methods, and not with the information being supplied. People get angry when they think you are belittling or mocking their beliefs, so you need to be polite and respectful. When we look at the stories of the people in this book who escaped the rabbit hole, we find that they did so because of finding some useful information they were missing. Willie discovered that Chemtrail tanks were just ballast barrels, Steve discovered contrails actually can persist, Richard discovered that heat weakens steel beams well before their melting point, Edward discovered that NIST (the National Institute of Standards and Technology) had a detailed explanation for the collapse of Building 7, and Stephanie discovered people promoting Chemtrails also wanted to shoot lasers at passenger planes.

All these people escaped the rabbit hole in part because of useful information they were missing. They were influenced by factual and logical information communicated to them with politeness and respect. They did resist initially and it's likely that some early attempts to communicate with them did backfire. But eventually the weight of the evidence ran its natural course.

Keep supplying useful information to your friend. Keep up the communication, the supplying of perspective, and the encouragement of a genuinely open mind. Resistance is an inevitable part of the process. Don't be disheartened. Give it time. It might take months, or even years.

Supplying Useful Information

Communication techniques can be discussed in the abstract, but supplying useful information is a very practical matter. What is the information that you should be supplying? The remainder of this book focuses on key examples of this information, and how to communicate them. We can identify a few types of information you should be supplying:

Identifying Errors—A simple explanation of something that is wrong is a core part of debunking, and one of the most common things that we will encounter. If someone thinks that the fires in the World Trade Center were nearly extinguished when it collapsed, then you can show them the photos taken by Greg Semendinger minutes before the collapse that show the North Tower ablaze with walls of flames 40 feet high and 200 feet wide.[8] It's a simple demonstration that they were wrong, presented without judgment.

Explanations—Demonstrating an error sometimes stands alone, but it works best if you can provide a more complete explanation for what is going on. This means discussing the underlying details, providing a real-world explanation that can replace the one you've shown to be false. This is best done with a practical illustration. For example, when they claim that hot spots under the World Trade Center rubble pile (weeks after the collapse) are evidence of incendiaries, then explain that incendiaries would have quickly burnt up *and* show them articles about landfill fires that burned underground for weeks,[9] months,[10] or even years.[11]

Exposing the source—We are often told that an *ad hominem* attack (an argument directed against an individual as opposed to the position propagated) is a logical fallacy. You cannot prove an argument to be false by attacking the character of the person making the argument. You should address the argument itself. This is a good general rule but there is an important situation where this does not apply—when your friend is using an *argument from authority.*

If your friend tells you that they believe 9/11 was a controlled demolition in part because "thousands" of professionals at AE911 Truth say it was, then it's perfectly reasonable to show that that AE911 Truth is composed of many non-experts, is frequently incorrect, and has historically repeated false information long after it has been proven false.

Make clear that this is not a personal or vindictive attack. Your friend has claimed that this organization is a reliable and authoritative source. You simply need to demonstrate that their source is not as reliable or authoritative as they thought. Demonstrating the errors the source has made, and the falsehoods they continue to repeat, is a great way of prompting your friend to take a more critical look at the source's other claims.

Perspective—Is the theory even plausible? How would it work in the real world? The reason that some conspiracy theories get traction is because people don't really have a good perspective on the context and the meaning of what is being claimed. Who would it actually benefit to fake the shooting of children? How much is a trillion dollars? How many tons of steel, concrete, and drywall were there in the World Trade Center? How much has air traffic increased since the 1980s? When did people discover the Earth was round? What percentage of structural engineers think that the World Trade Center fell because of a controlled demolition? How many millions of scientists are there that need to be silenced? How would that even work?

In the next part of the book we will look at many practical examples of communicating these types of missing information. We'll look at four major conspiracy theories that span the spectrum: Chemtrails, 9/11 Controlled Demolitions, False Flags, and Flat Earth. We'll meet people who've escaped from the rabbit hole in all four areas. But we'll start with someone who was down the rabbit hole for a long time, and then emerged to start his own campaign of respectful communication and supplying useful information to the people who used to reside down there with him.

CHAPTER SIX

Steve – A Journey through the Rabbit Hole

I met Steve online via Contrail Science, and we've met in person several times. He's an old-school conspiracy theorist who escaped the rabbit hole and now actively tries to help those he left behind. Steve's journey down the rabbit hole started long before Chemtrails and 9/11, with Watergate and aliens.

It started in 1973 with this guy I met at art school in Chicago, Sherman Skolnick. I was just working as a cameraman, and he was discussing how this Flight 553 that came from Washington to Chicago carrying Dorothy Hunt was brought down by Nixon's chief narcotics agent. He had all this evidence, like they found narcotics and high levels of cyanide in the pilot's blood. There was all this crazy stuff, he also mentioned that Spiro Agnew would resign, and months later he did. I didn't have any information that would counter that, this was before the internet. So it kind of put a little seed in my head—like the government is corrupt, they can do weird things behind the scenes. That's how it started.

Then I got into the whole Zecharia Sitchin ancient alien thing. I read all his books, it became fascinating, like entertainment. From there I went to David Icke and his first book on 9/11, Alice in Wonderland and the World Trade Center Disaster. *That was connecting all these background things of ancient aliens running this one world government behind the scenes, so it confirmed everything I'd accumulated reading Sitchin.*

Back then I was excited because I was one of the first ones to know this. It was novel esoteric information, and it increased my ego. It made me an interesting guy all of a sudden. I could talk about these ancient aliens and the connection with 9/11. So I kind of got into it like anyone else, partly for entertainment.

Steve's interest moved smoothly from Watergate conspiracies, to ancient aliens (the theory that aliens had a significant effect on early human development, like building the pyramids) and then to 9/11. He enjoyed the ego boost this knowledge gave him, the feeling of uniqueness. He then discovered another benefit, a community of friends.

I've read studies about how some conspiracy theorists are kind of lonely people. That's kind of what I've seen. Some of them are straight out outcasts, weird, you know, you've seen them. Then suddenly they've got a family. They have these meet-ups on the first of each month and there would be some really attractive ladies, and we'd have beers afterwards, and it became a family. Ever since then I've never had that many friends like that. We were working towards this goal, trying to change the world. Everyone would applaud if you made a good video. You'd stand up and give a little speech, and there would be this great camaraderie. Even the demonstration would be exciting, it seemed we were historical figures in the early onset of a revolution, waking up the world to this horrible government.

It was exciting, it was family, it was fun and for people who did not have much of a social life it fulfilled that too.

A lot of these people, because of their beliefs, they were alienated from their family and their traditional friends, so this became a substitute. They didn't consider it a cult, it was a political movement with goals to wake up the world, just like the patriots. When you are doing it you're elated, there's endorphins flowing. For me it was also all this literature I've been reading for all these years coming to fruition. I'm one of the leaders now. It's a big thing against the whole government and the aliens who rule our world. It's a fun, violence-free revolution.

We knew we were outcasts, we were doing something almost illegal, so it was kind of exciting. In the group we'd do survival techniques. We'd go up into Topanga Canyon and we'd practice as if the police or the military were after us, how would we survive in the woods. The group even went looking for FEMA camps [supposed government internment camps for an impending crackdown],

76

they found some suspicious water system or something. It was like we were about to take down the government, and that was really exciting.

The activities of the group seem to blur the line between a kind of fun fantasy role-playing and a dangerous descent into delusion-driven, almost violent, revolutionary action. It sounds like harmless games in the woods, but "looking for FEMA camps" is the type of thing that Timothy McVeigh did the year before killing 168 people in Oklahoma City.[1]

For someone whose conspiracy theories spanned the spectrum all the way up to aliens running the world, it was something more in the middle, Chemtrails, that became a dividing line for Steve.

The We Are Change people were very seriously into Chemtrails. I did this interview video, there were contrails making an "A" shape in the sky, and they were pointing to it, saying that's obviously a Chemtrail, a government plot. I went along, nodding my head, but didn't really believe it 100 percent. For me it just took looking up the word "contrail," and it said there were contrails that were short, and other contrails persisted and spread out, and it gave the reasons for that. I tried to explain that to the people at We Are Change, and they threw me out of the group temporarily. So that kind of undermined the whole cause of 9/11 Truth. I was trying to help them by explaining that Chemtrails were bullshit, but those guys were 100 percent behind the whole Chemtrail thing.

They were getting really intense, because they thought they were the first people to expose Chemtrails. This was right before the film What in the World Are They Spraying? *So, for me that was the impetus to start looking into it, and your site [Contrail Science] was just coming up so that was more information I used to defend myself. I nearly got in a fight with one guy because he thought I was the most horrible government agent because I didn't believe in Chemtrails. So that's pretty much when I left the organization.*

After Steve left We Are Change, he watched how it evolved. He continued to be friends with many of the people there and continued to attend meetings and demonstrations to try to explain what he had found out about "Chemtrails." But he observed some disturbing trends in the group, observations which helped him further out of the rabbit hole.

The whole group was originally somewhat intellectual, somewhat smart. But the guys who kind of took over were just weird. They would get into the whole other stuff, the Sandy Hook stuff [the idea that the school shooting at Sandy Hook was fake], and that was like filling up the whole conspiracy with dirt. They would have arguments because it was splitting up the group, then it became like a power thing. I couldn't get through to them, they would almost get to a point of violence with me when I would show up. So I decided to stay away.

Separated from the group, Steve decided to take a deeper look into the 9/11 conspiracy theories. He jumped in at the deep end with what Truthers consider to be their strongest evidence: Building 7.

After I left I got the courage to really look into the 9/11 stuff. I went to 911Myths, there were some mistakes in it, but it was pretty good. But really for me it was Building 7. I cut to the chase. I wanted to see what was the reason that building went down. There was the whole column 79 thing, there being no water for sprinklers, the damage to the side of the building, video and photos of the south side of Building 7, you could see the smoke pouring out of the windows, all the windows were broken. All that was never shown, never demonstrated to the Truthers, they just didn't even know that. Seeing that was a key moment for me.

I'd take the NIST slideshow and print it out and then go out to the We Are Change 9/11 demonstrations and try to explain it to them. At first, I was scared, because I was going against my friends. But then I got encouraged because people gathered around me because they hadn't seen it. And at that time they were getting forty to fifty people at these demonstrations, but after that it became very small, just a handful. So, I was having an effect.

Then I became almost a villain in the group. I went to one meeting at and one of the leaders, Katy, an actress, she was screaming on the bullhorn, "Look up, google Building 7," and I remember telling her right to her face: "I did, have you?" And she never came out to a demonstration again.

Another one happened during the Occupy Wall Street thing here in LA. Jeremy Roth was about to ambush Van Jones about 9/11. So, I said, "Jeremy, listen, you've got to look and understand it, the building collapsed due to kinetic force, the building floors, that was the energy source." I explained it to him as much as I could, and he didn't ambush the guy. I told him, you're a smart guy, you should be working towards your activism with real things. I never saw him again at any 9/11 demonstration. In fact, all four founders never went to another demonstration again. They just gave up.

Steve left the movement nearly a decade ago, and over the years has had plenty of time to reflect on what got him into the rabbit hole, what got him out, and how to help other people.

A lot of it is just laziness. You fall into it and you think you found something new, and since there's nobody arguing against it you get deeper and deeper into it. That's what happened to me, there wasn't any counterargument to 9/11, it was always just explained as government manipulating the media. It keeps self-confirming and you just keep getting more and more isolated.

Really the critical thing is just being able to have the courage. When [the group leaders] lose their credibility, it increases your courage to challenge what they are saying. 9/11 Trutherism is belief in authority. You think AE911 Truth is all the architects and engineers on the planet, when really it's just a handful, only a few who have any structural background.

I think a lot of 9/11 Truth is that they have this information they think is novel and nobody else knows about. If you have someone like me, an ex-Truther, I always tell them I already know everything you are saying to me, I used to do it myself, I used to do demonstrations myself. See, with you, Mick, and other debunkers like you, they just think that you don't know what they know. But with me I've already been down that rabbit hole and I know that information. I was there, I was a demonstrator, I was a soldier. I can tell them straight to their face that it's incorrect, they are not doing critical analysis, they are not looking at both sides.

They often just don't have the guts to look at the other side. They have invested so much, with family, with conversations at Thanksgivings, it's just very hard to go back against all your friends. So, part of the problem with getting out is just embarrassment, and you need the guts to get past that and actually look at the information.

You can give people information about their conspiracy theories and they won't take it. But you can give them info about other things that are obvious bullshit like Sandy Hook. Then the Chemtrail thing is excellent, as it's such a visual, you can see them every day. People get sucked in quick, but you can give them a bit of information about how contrails form, and they see that, and it gives them a bit of courage to question other things.

There's some people you definitely cannot get through to, there's just too much ego, too many Thanksgiving dinners they have invested in. There was one guy I was just explaining how claims have counterclaims, and there was this great process of critical thinking. I had been showing him the NIST 9/11 slides and he realized he was wrong at that moment. This guy just started screaming. It stuck in his head that he would have to go back to his family and explain that he was incorrect. He almost had a nervous breakdown.

So give them time, let them find the courage to look into it themselves. Let them look quietly in their room, on the internet, and then come back to me. I'd always challenge them: I know you're not going to believe me now, but just read the NIST report and then we can have an intellectual discussion of what you think is wrong with it. But until you do that we can't really have an intellectual rapport. I challenge them, not to believe it, but to be better informed for the next conversation. But usually they just never came back.

He still occasionally goes back to the local Los Angeles meetings, but it's not really the same. Demonstrations have tapered off, and most activity happens on the internet.

The only group here in Los Angeles meets on Saturdays at a café, and they have combined the 9/11 Truthers with the Chemtrail thing. But they are strange, it's a small group in a dark room. I've been there, but they hate my guts, the Chemtrail guys, so I'm a little nervous there. They only allow me one question, then they blow it off and go to someone else for an hour.

Once you are out, once you are no longer part of that group, of 9/11 or Chemtrails or whatever, you are kind of left empty. You're going to lose all your friends that you'd suddenly developed relationships with. There's some very shy introverted people, so the conspiracy group is all they have, and if they leave then the few friends they have in their life are going to be gone. It's very traumatic.

I'm proud of myself that I was able to get out of a cult. I feel like I flushed out that whole thing from my brain. You don't think you could belong to a cult when you are actually in a cult. You think you are too smart for that. But it slowly progresses around you until you are surrounded by it. It's amazing, how the ego works. Anybody can fall into the trap.

I'm grateful to you, and to all the people who contribute to Metabunk. Even though I don't write too much, I read it to get the latest information to make sure I'm up to date, in case I meet somebody who needs it.

Talking to Steve about his journey has been very useful to me. His experience is typical in some respects, but it's rare to get such a complete picture over such a long period of time. He was sucked in pre-YouTube via books as well as videos. The esoteric information the books and videos provided made him feel special, and then helped him find a community that would look up to him for his unique knowledge.

But events triggered a desire to look into a topic on the edge of what he believed, just over Steve's personal demarcation line. He was fully into 9/11, but then he looked into Chemtrails because others in his community believed that theory. Chemtrails turned out to be bogus, so his line shifted and solidified a bit. The people around him believed in both Chemtrails and 9/11 in greater depth, but he drew the line at Chemtrails. That pushed him away from the group and triggered the urge to look into 9/11 in greater depth. By then his self-directed journey out of the rabbit hole was well underway, helped by debunking sites like Contrail Science, 9/11 Myths, and Metabunk.

Not only did he get himself out, but since then he has helped many people out simply by communicating effectively with them, showing them useful information, exposing the problems with their leaders and sources of information, and giving it time.

PART TWO

CHAPTER SEVEN

Chemtrails

FIGURE 5: Condensation trails from commercial air traffic crisscross the sky near Sacramento, California. Chemtrail theorists think this is evidence of deliberate spraying.

The Chemtrails conspiracy theory claims that some of the white trails left behind planes high in the sky are actually being deliberately sprayed for nefarious purposes. In popular discussion of the topics (by non-believers), the theory is often described as one involving spraying some kind of mind-control poison. But that's really a fringe version of the theory, not believed by many. The most common explanation given by most Chemtrail believers is that the trails are part of some kind of plot to control the weather and/or the climate.

More specifically, the most common versions of the theory claim that there is a secret program of climate modification called "geoengineering." In this version of the theory, the long-lasting trails left by planes are some kind of substance that blocks out the Sun and are designed to cool down the planet to counteract the effects of global warming.

This theory is to some degree based on very real science. Geoengineering is a real subject with reputable scientists researching it. Going back well over fifty years, there have been numerous proposals for methods of cooling the planet by spraying chemicals into the air. But none of these proposals have ever made it past the theoretical research stage. The only significant experiments that have been done have been simulations on a computer—kind of like an advanced weather forecast. This is for good reason—engineering the climate to reverse the effects of global warming is not something to be entered into lightly. It is something that would have to be done on a planet-wide scale because the climate is a global system and local changes quickly spread out across the entire planet. It's also something where we don't really know what the effects would be, or how to safely stop it once started.

Suppose that we gave it a go, only to find that it drastically reduced rainfall in China causing crop failure, widespread famine, the deaths of millions, the collapse of the Chinese government, and followed quickly by global economic crises, war, etc. Conversely, it might cause massive monsoonal flooding in India and Bangladesh, with similar catastrophic and destabilizing results.

Given these genuine and significant concerns, it's clear why nobody would secretly be carrying out a program of geoengineering. Why would they, if they don't know what the result would be? Why would they do it if the treatment had an unknown probability of being far worse than the disease?

This argument against the "covert geoengineering" theory is especially valid if we consider how old the Chemtrail theory is. The first version of the theory arose in 1997 among the more typical conspiracy theorists at the time—anti-globalists concerned about a "New World Order" who thought there was a depopulation plan involving spraying chemicals to make people sick.[1] This theory quickly morphed into the geoengineering theory with the writings of William Thomas. Consider what the age of the theory means for your friend—exactly what could they have been spraying twenty years ago if even now the most advanced geoengineering scientists don't know how to manage the potentially cataclysmic side effects of geoengineering? If we don't know *now* what the side effects might be, then how on earth could anyone have thought it a good idea two decades ago?

As well as debunking all the claims of evidence, which we shall do later in this chapter, a key part of helping your friend out of the Chemtrails rabbit hole is supplying them with genuine information about the true state of geoengineering research. While a full review of the scientific literature would be ideal, it's generally not practical. Focus on the state of geoengineering field experiments—tests that have actually been done in the wild, spraying things from planes to see what a particular geoengineering technique will do.

It turns out that there have been practically zero such tests (at the time of this writing in 2018). In 2009 Russian scientist Yuri Izrael sprayed some smoke out of a helicopter at around 650 feet above ground see how much it blocked the Sun.[2] The most well-known geoengineering researcher in the world, David Keith, described this as nothing more than a publicity stunt, as it did not really replicate anything like geoengineering. The second test (and so far the last, and arguably only) was in 2011 by the Scripps Institute of Oceanography, in which smoke from ships and salt from planes was deliberately sprayed on clouds at low altitudes to see how it changed them.[3]

Besides those two events, there's still only been a handful of proposals for field experiments. In 2013, Keith and others published a paper that discussed various proposal for possible tests.[4] In 2014, a test was proposed using a tethered balloon to do the spraying.[5] Planning was started for this test (the SPICE test), but it never happened due to a variety of objections.

Keith hopes to try again in 2018, with a variety of tests. The previously proposed balloon experiments will hopefully take place. That should give the researchers an idea of whether tethered balloons are a viable method. In addition, a variety of substances (sulfur dioxide, aluminum oxide, and calcium carbonate) will be sprayed from planes at high altitude. In each instance less than a kilogram (two pounds) of each material will be sprayed.[6]

Not only have there been no real field experiments to date, but there's still considerable discussion whether geoengineering will even work, which method to use, what the side effects will be, and what will happen if we do it and then have to stop. A 2016 analysis of the state of geoengineering research carried out by the American Geophysical Union listed twelve different areas of uncertainty that still needed to be resolved and concluded:

> *Any well-informed future decision on whether and how to deploy solar geoengineering requires balancing the impacts (both intended and unintended) of intervening in the climate against the impacts of not doing so. Despite tremendous progress in the last decade, the current state of knowledge remains insufficient to support an assessment of this balance, even for stratospheric aerosol geoengineering (SAG), arguably the best understood (practical) geoengineering method.*
>
> *Few would assert that the current state of knowledge is sufficient to support a decision to deploy. Ultimately, any decision to deploy decades-hence will necessarily involve a tradeoff confounded by risks associated with either choosing deployment or choosing not to. . . . Research should thus aim to address key uncertainties associated with SAG, such as those listed here, in order to support well-informed future decisions. Up to now, geoengineering research has been dominated by scientific questions, but as research proceeds, it will need to also address important outstanding engineering or design questions. Instead of asking, "What will geoengineering do?" we will have to ask, "Can geoengineering do what we want it to do, and with what confidence?"[7]*

If your friend thinks that secret geoengineering is an obvious explanation for the "Chemtrails," then you should try to supply him with as much information as possible about the actual state of geoengineering research. Explain just how uncertain the world's preeminent scientists in the field are *now* about the effects and effectiveness of geoengineering. Explain how the theory started in 1997, and yet we *still* have not gotten around to doing any high-altitude field experiments. Explain how the most recent assessments still discuss it as something we might do "decades hence."

But as well as looking at the implausibility of the theory and the evidence against it, we also have to look at the claimed evidence *for* the theory. Like all mature conspiracy theories, there's a broad spectrum of variants of the Chemtrail theory. These in turn are based upon a varied set of claims of evidence. As with

9/11, there's a core of these varied beliefs and claims of evidence, and there's a single website that is most frequently referenced by 90 percent of believers. With 9/11 it's *Architects and Engineers for 9/11 Truth*, run by Richard Gage, and with Chemtrails it's *Geoengineering Watch*, run by Dane Wigington.

While looking at the more esoteric variants of any theory can be an important part of helping your friend out of the rabbit hole, in almost every case a significant portion of their belief is going to be based around a relatively simple core set of beliefs. This is especially true with Chemtrails. When attempting to understand and help someone who is deep down the Chemtrail rabbit hole, the most productive approach is to directly address these core claims of evidence. These core beliefs are all brought up in the core website—Geoengineering Watch.

This concentration of evidence is very useful in the "consider the source" aspect of debunking. People like Gage and Wigington are believed because of their good nature and seeming expertise. If you can demonstrate conclusively to your friend that the core promoters were wrong about one particular claim of evidence, then this allows your friend some mental freedom to begin to question other claims of evidence. It also gives you more leeway in questioning the claims. You can directly ask your friend: "If they are wrong about A, is it possible they are wrong about B?" Remember that this is not meant to entirely discredit the other claims by association, rather to raise the need for some fresh scrutiny of those claims.

I'll briefly list the core claims of Chemtrail evidence here, and then go into each one in more detail and discuss how to approach the topic, and how to deal with common rejections and objections to the facts. The claims of evidence are presented roughly in the order of likelihood of your friend bringing them up.

Contrails Do Not Persist—By far the most common misconception underlying the theory is that since contrails are just condensation, then they should always quickly evaporate (like your breath would on a cold day). In reality, contrails are a type of cloud, so they can do whatever clouds do, including persist. We also have extensive historical evidence of contrails persisting (sometimes for hours) in decades' worth of books about clouds and weather, as well as extensive collections of old photographs.

Weather Modification Exists—Weather modification (a.k.a. cloud seeding) is a technique for increasing rainfall by spraying things into a cloud to prompt the formation of raindrops. Many Chemtrail theorists have simply never heard of it, so when they first come across an actual example of cloud seeding they think that it proves the Chemtrail theory. But cloud seeding has been done quite openly for sixty years, it's done with small planes on existing clouds, and it does not leave a trail.

Videos Show Chemtrails Being Sprayed—Sometimes described as "irrefutable proof" of Chemtrails, there are several videos of planes leaving a type of contrail called an "aerodynamic contrail" that forms over the wing, and not from the engines. The misidentification of these types of trails as a kind of spraying appears to be based purely on a lack of knowledge of this type of contrail. The best approach here is to try to fill in the gaps in their knowledge.

There Are Photos of the Inside of Chemtrail Planes—These are either photos of pre-production planes with ballast barrels, or other known planes with tanks inside of them, like firefighting planes. All of the photos presented as evidence of Chemtrails have been verifiably explained as something else.

Chemical Tests Prove Chemtrails—Here your friend will show you a variety of chemical analyses of air, water, or soil that they say show unusually high levels of some chemical element (often aluminum, barium, and sometimes strontium). These tests are almost invariably the result of misunderstanding how much of these elements naturally exist (soil in California, for example, is 8 percent aluminum on average). They are also often the result of poor sample collection techniques, or just misreading the units in the results (confusing milliliters with microliters).

Modern Jet Engines Can't Leave Contrails—Most large planes use a type of engine called a high bypass jet engine. Older (low bypass) engines got their thrust (the force that pushes the plane forward) largely by shooting out the jet exhaust from the back of the engine. Newer high bypass engines use the power of the jet engine turbine to turn a very large fan at the front of the engine, and a much larger portion of the thrust comes from the air pushed by that fan. Chemtrail theorists claim that this means that it can't leave contrails. The problem here is that contrails are made from the exhaust gases of an engine, and not from the thrust. Both high bypass and low bypass engines burn fuel the same way, so they both leave contrails. Conveying this can be a bit of a challenge.

Patents Prove Chemtrails—The argument here is that since patents for geoengineering, weather modification, and spraying things out of planes exist, then that proves there's a covert geoengineering program using Chemtrails currently in effect. They will give you a long list of patent numbers and names. The information they are missing here is that there are many things for which patents have been issued and which do not exist (for example, when was the last time you saw a space elevator?). There's also quite a lot of patents on their list that really have nothing to do with geoengineering, or even spraying.

Photos Show One Plane with a Contrail and One with a Chemtrail—This is something you might see for yourself quite often if you watch planes for a while. One plane is leaving a tiny non-persistent trail (or even no trail at all) while another plane, seemingly at the same height, leaves a thick persistent trail that spreads out. The missing information here is how much the atmosphere varies in just a few thousand feet (even a few hundred). Contrail persistence is quite sensitive to small changes, like how water freezes at 32°F, but not at 33°F, and so planes very close together can leave different contrails.

The Government/UN/NASA/CIA Has Admitted "It"—there's a variety of reasons given for this belief in admissions, but it's generally a misunderstanding of either research into possible future geoengineering, or unrelated things like weather modification or space sounding rockets (rockets carrying scientific instruments for research purposes).

There are other claims of evidence that are made by Geoengineering Watch and others. For example, there are claims that UV levels are blindingly high, or that the trails are the wrong color, or that planes fly in odd patterns to alter the atmosphere so HAARP can cause earthquakes. These claims have been addressed on sites like Metabunk, and you can find the explanation there with a bit of googling. But it's best to try to put the more esoteric claims aside for now. They largely rely on unreliable information or subjective assessments that in turn are only given credence because your friend has a tendency to believe in the promoter of the theory. If you can get through the explanations of some of the more fundamental core beliefs then it will make handling the more esoteric ones more straightforward, since you will have removed that tendency to blindly believe.

Contrail Science

To understand some of the problems with the false claims behind Chemtrails, it's helpful to have a basic understanding of the science behind contrails. I will keep this as simple as possible, and there's plenty of other resources to find more detail.

There's one simple fact that's related to the vast majority of confusion about contrails (and hence the Chemtrail theory), and that is: *contrails are a type of cloud.* Contrails have been classified as clouds by the World Meteorological Organization for decades. Like other clouds, contrails have now been included in the Latin taxonomy of clouds, informally as *cirrus aviaticus* (aviation cirrus) and more formally *cirrus homogenitus* (man-made cirrus). Contrails are man-made clouds, like the clouds sometimes found over power stations (*cumulus mediocris homogenitus*), but they are still clouds nonetheless.

What is a cloud? People often only have a loose idea and think of clouds as "water vapor." But water vapor is actually an invisible gas. Clouds are made of trillions of tiny drops of *liquid* water suspended like fine specks of dust in the air. If the clouds are cold enough (usually very high up in the sky) then they are made of trillions of tiny ice crystals.

When people tell you that a contrail should quickly dissipate, they usually refer to it as "condensation." They compare this to the condensation of your breath on a cold day. Here they are partly correct. Contrails from jet engines form because hot jet exhaust has water vapor in it, just like how your breath is warm and humid. When this hot, humid breath (or jet exhaust) hits colder air, then the low temperature makes the water vapor condense. It condenses into a cloud, physically not much different to a tiny part of a typical cumulus cloud you might see in the sky.

But the condensation cloud from your breath quickly vanishes. This is because in order for a cloud to persist the air needs to be humid enough. For a liquid water droplet cloud to persist it needs a relative humidity of 100 percent, e.g., inside a region of cloud or fog. For your breath to form clouds that persist you would essentially have to already be standing inside of a cloud, and your breath would just be adding a little more condensation to it.

Why then do contrails (sometimes) persist in a clear blue sky? If contrails are condensation clouds like your breath, then why don't they vanish almost instantly like your breath vanishes? It's because contrails freeze.

Contrails from engine exhaust start out exactly the same as the condensation from your breath, as a cloud of billions of micro-droplets of water. But since the temperature is very cold in the upper atmosphere (-40°C/-40°F), the droplets freeze. Once frozen they cannot evaporate. *A contrail is the frozen breath of a jet engine.*

If the humidity is very low the cloud won't last very long, even frozen ice will fade away at very low humidity (a process called "sublimation," where a solid goes directly to the gas form). But the level of humidity needed for ice clouds to persist is *much* lower than the level needed for water clouds. Water clouds need 100 percent relative humidity with respect to water, but ice clouds need only 50 percent to 70 percent (the actual amount decreases with temperature).[8]

This explains why we get contrails when there's hardly another cloud in the sky. Ice clouds (like cirrus clouds) need a higher humidity to form than they need to persist. The entire sky can be just ripe for clouds to live in, but they just need a little bump of moisture to get them to spring into existence. The water in the jet exhaust provides that little bump, popping the humidity up high enough for water vapor to condense, quickly freeze, and then form ice clouds. These ice clouds are contrails, and if humidity is high enough they are *persistent contrails*.

Concepts such as relative humidity and trillions of micro-drops instantly freezing can feel rather alien and incomprehensible. Don't worry too much if it takes you a while to get used to them. Conversely, if you are already familiar with the science, or if it all seems quite straightforward to you, then keep in mind that it almost certainly will not seem that way to your friend. Give it time. It can help to show a variety of different versions of the explanation (like in the books we shall discuss in the next section).

Another key concept to understand is that clouds form in two ways; the most common is by warm air rising until it cools below its dew point. Most clouds you see in the sky form this way, and if you look at a time-lapse of clouds forming you'll typically see them seeming to boil upwards.

A less common type of formation is the "mixing cloud;" this is formed when two volumes of air with different temperatures and humidities mix together, and the resulting mixture is cool and humid enough for a cloud to form in. Exhaust contrails are mixing clouds; they form from the mixing of humid exhaust gases with outside air. In the glossary of the American Meteorological Society, the definition of mixing clouds says, "An example of a mixing cloud is a condensation trail."[9]

Finally, and crucially, there's two types of contrails. The contrails described thus far, where exhaust gases mix with outside air, are called "exhaust contrails." The other type is called an "aerodynamic contrail."

I cannot stress strongly enough how important knowledge of aerodynamic contrails is to the effective debunking of the Chemtrail theory. There's many videos and photos that chemtrail believers present as "irrefutable proof" of spraying that are simply aerodynamic contrails. Even if you don't yet understand the science behind how exhaust contrails work, you still need to understand that these aerodynamic contrails exist and that they are different from exhaust contrails.

You've probably seen a type of aerodynamic contrails if you've ever landed in a plane in damp weather. Condensation clouds form over the wing because the air flowing over the top of the wing drops in pressure. If it's humid enough this pressure drop can trigger condensation. Usually when you are landing, any clouds that form on top of the wing almost instantly dissipate when the pressure returns to normal. Sometimes you see long cylindrical streamers of cloud coming from points on the wing—these are vortices formed by spinning air which keeps the pressure low and the aerodynamic contrail visible for much longer—but when the air stops spinning, the contrail fades away just like it did over the wing. Hence, aerodynamic contrails are short-lived at ground level.

The story is very different at high altitude. If it is cold enough then the drop in pressure over the wing will cause the water vapor in the air to both condense from vapor into tiny liquid water droplets and then freeze solid into tiny ice crystals. This process fixes the aerodynamic contrail in the air (by freezing) in the same way an exhaust contrail is fixed. But because these aerodynamic contrails follow a different mechanism of formation, the atmospheric conditions required to form are also different. You still need the humidity, but because the temperature drop is caused by the pressure reduction over the plane's wing, you don't need the air to be so cold. Because of this, aerodynamic contrails can form at much lower altitude than exhaust contrails. Exhaust contrails typically start at around thirty thousand feet over the continental United States, but aerodynamic contrails can form as low as twenty thousand feet, or even lower in cold weather.

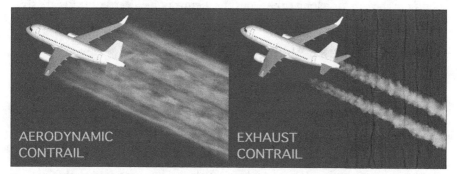

AERODYNAMIC
CONTRAIL

EXHAUST
CONTRAIL

FIGURE 6: There are two type of contrails: Aerodynamic and Exhaust. Chemtrail theorists often seem unaware of aerodynamic contrails.

The different types of contrails also look radically different close up. You've probably seen lots of them in the sky and not realized they were not coming from the engines. With a bit of practice you can tell the difference from a distance—aerodynamic contrails usually look flatter, more ribbon-like. But the real difference is close up. Exhaust contrails form directly at points behind the engines, with a gap (of varying size) between the engine and the trails. There's one contrail per engine so you typically will observe two or four (or occasionally one or three) quite distinct trails that eventually merge. By contrast, persistent aerodynamic contrails form along the entire wing, and therefore start out as a single wide sheet. Different shaped parts of the wing have slightly different pressure effects, which can result in multiple stripes forming across the trail. These stripes are sometimes interpreted by chemtrailers as spray coming from nozzles, but it's just variations in the trail coming from the entire wing. The trail forms in a thin layer and the condensing water particles sometimes interact with the Sun creating odd rainbow colors. These colors are sometimes interpreted by chemtrailers as evidence of chemicals being sprayed, but it's just pure water from the atmosphere. It's all ripe for misunderstanding if you don't know what you are looking at.

Contrail Persistence

The root claim of evidence, the foundational belief that nearly all Chemtrail believers have, is that normal contrails cannot persist. In reality, contrails are a type of cloud, so they do just whatever clouds do, which obviously includes staying in the sky for a long time. For the Chemtrail believer, contrails are simply "condensation" and therefore should quickly evaporate. The misconception can be traced back to the writing of William Thomas, who said in 1999:

> Experienced pilots, military personnel and other qualified observers note that . . . normal contrails dissipate less than one minute after formation.[10]

He did not provide the names of any of these qualified observers. He's actually flat wrong, but the claim somehow stuck and has persisted to this day.

The very first thing to do when talking to your friend about their Chemtrail belief is to directly address this false claim. You could start out by asking them how they know that it's true, how did they discover this fact, did they read it in a book? This question tends to lead to some circular thinking—they know the Chemtrails persist because when they see Chemtrails in the sky they persist, whereas the contrails fade away. It's always worth a shot, and if they seem receptive and want to explain how they know, then that's a good opener. Understanding where they got one false idea gives you a great follow-up for when you explain that false idea. What other ideas did they draw from the same source? Maybe some of those ideas might not be as solid as they once thought. Here we're employing the 'exposing the source' method described in Part 1 as a means of supplying useful information, and beginning to sow doubt in a unreliable source.

Another potential approach is to discuss the science of contrail persistence. Ideally one should be able to point out relatively simple things from the previous section (like contrails being clouds, so why can't they last as long as clouds do, or the temperature being well below freezing so why do they think contrails would evaporate instead of freezing?), but this approach can be tricky. Unless they are willing to learn a few new concepts (like ice supersaturation and homogenous ice nucleation, for example) then this approach can lead to confusion. If they can't grasp the concepts, then it's going to boil down to you asking them to simply trust you. Diving into the science might not be the best thing to open with—especially if you yourself don't have a firm grasp.

An effective approach is to show your friend that historically contrails have persisted. There's a number of ways you can do this. You can show them old photos dating back to World War II where contrails are seen to be persisting and spreading. Since the Chemtrail theory dates back to 1997, I created a thread on Metabunk titled "Pre–1995 Persistent Contrail Archive,"[11] which contains approximately a thousand images and videos predating 1995 (going back as far as the 1930s).

The thread (which started in 2012 and is still expanding to this day) contains a wide variety of images. These include people's personal photo albums from the seventies and eighties that have been scanned and uploaded to sites like Flickr. Another very effective type of evidence for persuading your friend is images of persisting contrails in older movies and TV shows like *Terminator* (1991), *Irma La Douce* (1963), and *Spartacus* (1960). Since these are films that your friend may have watched before, and can certainly watch again, they are forced to either admit that the contrails were in the old films, or that somehow the films were all remastered later to add the contrails.

A thousand images of old contrails from hundreds of different sources should be enough to sway them somewhat. But the most compelling evidence comes from a relatively small subset of that collection—the images and descriptions of persistent contrails in old books on clouds.

When I first started to write about contrails (and the Chemtrail theory) I used resources like NASA.gov and current scientific papers as references. The problem with that is that anything post-2000 is suspected by the conspiracy theorists as being part of the cover-up. I started to use older books that I'd found scanned on Google Books. I was then told that since they were on the internet they were probably forged, again as "part of the cover-up."

I went analog. Ignoring the internet, I started to acquire physical books on the weather, anything that described how clouds worked. I found that invariably any weather book would include a section on clouds, and there were also many books that were entirely devoted to clouds. These books varied from simple collections of pretty photos to more involved discussions of the physics of the weather. But something that almost all of them had in common was photos of contrails that looked pretty darn persistent, often accompanied by a description of the varieties of contrail (persistent, non-persistent, persistent spreading, exhaust, and aerodynamic). Often these discussions would provide explanations for factors that the Chemtrail people find suspicious, like gaps in the contrail and contrails with rainbow colors.

Eventually I collected over thirty books on the weather and clouds. Many of them are available for a few dollars from secondhand booksellers over the internet, using sites like Amazon.com, eBay.com, and alibris.com. Some I found in physical bookstores or in library sales. Since these books are all at least a few decades old, many have been supplanted by newer versions and may be outdated. But what I wanted to use them for was to show people actual, physical proof that, like they still do today, contrails were recorded as persisting over the course of many decades.

That turned out to be seven decades. It's nearly ten decades if you look at online scans of published written accounts of persistent contrails, which go back to the early 1920s. The oldest book I managed to procure a physical printed copy of was published in 1943. In *Cloud Reading for Pilots*, page 73 reads:

There is one other form of ice crystal cloud. The artificial one caused by aircraft, and there are two varieties. One is produced by the condensation and freezing of exhaust vapour, which on issuing [from the engines] expands to a large extent forming a line of cloud which drifts in the sky for some time, spreading outward, and becoming less dense.

This was accompanied by two photos of contrails, the caption to one photo noted that it had a gap in it (something that Chemtrail theorists also claim is impossible in a "normal" contrail), and explained:

The broken part of the line is probably due to some local change in the [atmospheric] layer concerned.

All of this was contained within a delightful little book, a first edition that's seventy-five years old, and which I got for just ten dollars. There's no doubt to the item's authenticity when you see it, bound in fraying and fading blue cloth, the paper slightly yellowed with age, the whiff of musty ink and paper rising from its pages. Seeing this book, we know that A. C. Douglas wrote it in 1943, and that in 1943 experts in the fields of meteorology and cloud formation knew that contrails could persist and spread, and that they could have gaps in them.

If your friend is especially suspicious then he or she might claim your old book to be an incredibly detailed forgery. It's helpful to look at as many different books as you can. I created a video of my collection.[12] I start out telling Chemtrail believers not to trust some guy on the internet (i.e., me). Then I filmed myself opening each book in turn, showing the old contrail photos, and reading the paragraph that described how contrails could sometimes persist and spread. I started with a book from 2002 and went back to the 1950s (the oldest books I had at the time of filming). The whole video is less than four minutes long.

This video has proven to be one of the most useful Chemtrail debunking tools ever created. Many people who were once deep down the Chemtrail rabbit hole have told me that this was the video that started them on the path out. In light of the overwhelmingly positive response, I highly recommend that it be one of the first things you show your friend.

There's a curious disconnect here between some of the most vocal Chemtrail promoters and the vast majority of regular folk who happen to believe in Chemtrails. Most regular Chemtrail believers think that contrails do not persist. They generally think this is so because of the way they first experienced "Chemtrails." They either noticed some persistent trails and then while looking up an explanation came across the Chemtrail theory, or they heard about the Chemtrail theory (perhaps from a video or a radio talk show) and then a day or so later noticed persistent trails. Either way they did not remember seeing them before. Because of this, they accept the idea that the persistent trails are a relatively new thing, and that normal contrails should not persist.

The Chemtrail promoters have been arguing about Chemtrails for much longer and have had to eventually consider objections to their theory. Since the case for persistent contrails is inarguable, they have had to modify their theory to a weaker form where contrails shouldn't persist quite as often as is observed, or they have to double down into an unsupportable version where the Chemtrail plot has been going on since the 1920s, and consequently there must be a vast Orwellian conspiracy in which all the books ever written everywhere in the world are simply *pretending* that contrails persist.

Part of the Metabunk debunking technique is to demonstrate where the promoters and supposed "experts" are wrong. We can't always do this with contrail persistence as it's something the most vocal promoters have moved away from. If your friend is not entirely convinced by the fact that contrails actually do persist sometimes, then they are going to move where the promoters moved, to geoengineering patents, chemical tests, photos of barrels, videos of spraying, and the strange theory of high bypass jet engines.

High Bypass Jet Engines

The slow realization by the Chemtrail promoters that contrails *did* actually persist created a kind of cognitive dissonance. After all, Chemtrail promoters had always believed that contrails quickly faded away, and then they saw old books, photos, and video, showing them without a doubt that there are many decades of records and explanations of contrail persistence. Their theory relied on contrails not persisting, so was their entire theory wrong?

Some people probably did realize at that point that the theory was actually incorrect, and eventually this led to them escaping the rabbit hole. But those that remained managed to incorporate the persistence of contrails into their theory by a stunning display of doublethink. Historically, contrails *did* persist, as evidenced by the old books and photos, but in modern times, they claim, *contrails cannot persist* because jet engines are different now. Not only that, but they also go so far as to say modern jet engines can *never* make contrails.

For reasons we'll get into, this is flat wrong. Modern jet engines actually make contrails more often than the older ones. But before we get into why, it's very important to realize that this is a very seriously made claim that has been promoted by sites like Geoengineering Watch for several years. The fact that this claim has been disproven, and demonstrably so, is a very valuable tool in getting your friend to think about why they trust the claims made by those sites. Shining a spotlight on a single topic like this can be pivotal, as it was for Willie with the ballast barrels.

One recent (2015) article on this topic, titled "High Bypass Turbofan Jet Engines, Geoengineering, and the Contrail Lie," states:

They say that it is perfectly normal for this "condensation" to stay in the sky for hours or days, widening and spreading until whole horizons are completely blotted out. . . . Here is the fact of the matter, all commercial jet aircraft and all military tankers are fitted with a type of jet engine that is by design nearly incapable of producing any condensation trail except under the most extreme circumstances, the high bypass turbofan.[13]

Jet engines are basically long tubes: air is compressed in the front of the engine (the compressor), then that compressed air and the jet fuel are ignited in the middle of the engine (the combustion chamber), and finally the byproduct passes through the back of the engine (the turbine), which recycles some of the pressure of the burnt gases to power the front compressor. The hot high-pressure exhaust is then shot out the back to provide thrust.

This simple type of jet engine is powerful, but not very fuel efficient and is typically only found on fighter jets. Most passenger jets today have an enlarged fan on the front of the engine; the turbine turns this fan which pushes air into the compressor, but also pushes air *past* the engine to provide part of the thrust. This makes the engine more fuel efficient. The air that is pushed *past* the engine is called the *bypass air*.

The bigger the fan, the more bypass air there is, and the more thrust comes from this bypass air. Fighter jet engines have no extra fan at the front, so are referred to as *zero bypass*. Older engines, like those used on a Boeing 707, were very low bypass engines. Newer engines such as those you might see on a Boeing 737 or newer are almost all *high bypass* engines, simply because they are much more fuel efficient.

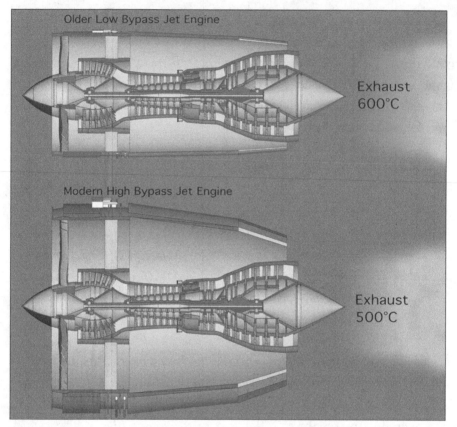

FIGURE 7: Older low bypass engines and newer high bypass engines share the same core, and both make contrails. The cooler exhaust of a modern high bypass engine creates more contrails.

The claim that newer, high bypass jet engines can't produce contrails, which Geoengineering Watch came up with, is best encapsulated by this narration from a 2015 YouTube video by Dane Wigington, creator of Geoengineering Watch:

Eighty percent of the air that passes through a high bypass turbofan jet engine is non-combusted. These engines are in essence a jet powered fan. The amount of non-combusted air passing through a high bypass jet engine is far too high to be conducive for any condensation formation. The high bypass turbofan jet engine is, by design, incapable of creating any condensation trails except under the rarest and most extreme circumstances, and even then, any trail would be nearly transparent and only momentary.[14]

Wigington's claim is wrong because contrails are not made by the bypass air, so it's irrelevant how much bypass air there is. As mentioned, contrails are made by the exhaust gases that come out of the combustion chamber and the turbine.

Jet engines require a certain mix of air and fuel to burn efficiently. This mixture in the combustion chamber is essentially the same in both high bypass and low bypass engines. Since the mixture is the same, that means the composition of the exhaust is the same for both types of engine.

An exhaust contrail forms when the exhaust gases mixes with the ambient air. It does not matter much if that mixing happens with the bypass air pushed past the engine with the fan or if the mixing happens with the air that's simply rushing past the engine from the motion of the plane. The end result is the same: the hot and humid exhaust gases mix with cold air and condense, then freeze, and contrails are formed.

Wigington and others seem to have got the idea that the bypass air somehow dilutes the exhaust gas. They miss the fact that dilution (mixing) is what *causes* the contrail in the first place. Contrails are mixing clouds, that's how they work.

Not only is the basic exhaust contrail formation unchanged between high and low bypass engines, there's another twist. Using the exhaust to turn the larger fan extracts some of the energy generated from the exhaust, which cools it down. Cooler exhaust reaches the condensation point quicker when mixing with the outside air, and so is actually *more* likely to make contrails than the hotter exhaust found in low bypass engines. There is extensive scientific literature on this, dating back many decades.[15]

There's a number of things you can discuss with your friend here. Firstly, there's the sheer scale of what is being claimed. If, in fact, new jet planes could not make contrails, then that would mean that *every single commercial jet* in the entire world has been created with this secret spraying program in mind. It means that almost every single time you see a contrail it's actually a "Chemtrail." This means that all the pilots in the world are in on the conspiracy, also all the ground crews, all the engine manufacturers, the FAA, and everyone in government, the entirety of the FBI, the CIA, the KGB, and all the corresponding organizations around the world.

The simple fact here is that Wigington, and all the other Chemtrail promoters who follow him, are not only wrong, but they have it completely backwards. *High bypass jet engines make more contrails, not fewer.*

It can be extraordinarily difficult to get your friend to accept this information. It can depend on how comfortable they are with the science. It can be helpful to look at old books that describe how contrails form. Try to work with them to understand how a jet engine works, what air is coming in, and what comes out. Find the common ground with your friend over how clouds generally are formed; then you can move on to the larger hurdle of convincing them contrails are actually just clouds; and so on. Focus on the fact that it's the *mixing* of the exhaust with other air that actually makes the contrail, so mixing with the bypass air will still make a contrail. The dilution doesn't make anything less likely: dilution is mixing, *mixing makes contrails.*

Remember it might take a while. Give your friend time to absorb the concepts. Don't push it. It's something that can take as long as is needed—the last thing you want is for them to stop listening to you.

But when they do, finally, hopefully, accept that high bypass engines are just as capable of making contrails as low bypass engines, then you've got a wedge. The fact that Geoengineering Watch (or whichever promoter initially hooked them on Chemtrails) got this wrong years ago, plus the fact that they have not corrected it despite being shown overwhelming evidence they were wrong, should be something that will make your friend start to question other claims made by that promoter. Perhaps when the promoter says they have "irrefutable proof" of Chemtrail spraying, your friend will consider that the promoter also said that high bypass jet engines were irrefutable proof. They are not.

Chemtrail Videos

One of the more frequent claims made by Geoengineering Watch is that they have "irrefutable proof" that planes are spraying Chemtrails. This supposed proof takes the form of videos that apparently show planes spraying. Almost invariably though the video shows a plane that leaves the type of contrail that they seem to be unaware exists, the aerodynamic contrail. It's basically a giant misunderstanding.

This misunderstanding dates back to a hoax carried out in 2010 when a video was posted by an Italian Chemtrail promoter, Rosario Marcianò, who goes by the name "Tanker Enemy" on YouTube.[16] The video shows a three-engine plane (a KC-10, US military) filmed from another plane that's following behind and above. Condensation is seen streaming from the wings, and the sunlight is shining on it in a way that creates a rainbow-like iridescent array of colors behind the plane. The footage is accompanied by various annotations that identify "nozzles" that are supposedly doing the spraying. Most amusingly though, is the audio from the pilot and flight engineer in the plane with the camera:

ENGINEER: You see them spraying that Chemtrail?

PILOT: Yup. Good thing we are above it.

ENGINEER: I know!

PILOT: Because we'd be dead right now.

ENGINEER: I gotta put this on YouTube.

PILOT: *[laughs]* Are you videoing right now? Oh God, don't video right now!

ENGINEER: It's like it's spraying out of the top of its wings.

PILOT: Don't do any evidence!

We now know that this is not a "Chemtrail" for several reasons. The first is that it's recognizable as an aerodynamic contrail. You can clearly see that it's not coming out of "nozzles," but is instead forming, seemingly out of thin air, over the surface of the wings. The best way to convey this to your friend is to show them information about aerodynamic contrails. See Figure 6, and do a Google image search for "aerodynamic contrails" and you will find many images of planes leaving very similar contrails, including the iridescent colors. In many of these photos it is very clear that the trail is coming from the entire wing surface and starts out invisible—which means it can *only* be water condensing out of the air. Chemicals being sprayed would come from specific points on the wing, and would start out as thick, visible trails.

The second reason we can tell this is a hoax is the dialogue between the pilot and the engineer. Any natural English-speaking person should be able to tell that they are simply joking. The idea that an actual member of this lethal and illegal Chemtrail-spraying program would not only joke about being "dead right now," but also joke about putting the video on YouTube, and *then actually do it*, is just silly.

Either of these reasons is sufficient to discount this video as evidence, but the case was sealed even tighter by the publication in 2011 of the original video by the engineer who took it, Tim M (a.k.a. USAFFEKC10). Tim posted the unaltered full-length video,[17] and later posted some still photos he took at the time that proved conclusively that he was there.[18] He wrote:

> *This is the original unadulterated video that started all the fuss. It is completely authentic, and no camera tricks were used. It is simply a couple of KC-10's in formation and the audio you hear is just us poking fun at all the "Chemtrail" [conspiracists]. I knew when I shot the video that this would be catnip for all the [conspiracists] out there. Yeah, the contrails have an odd way of "starting" and "stopping" but that is easily explained with physics. It's no different than the lenticular clouds that form over a mountain or the fog that flows from an open freezer. So, stop being so gullible, kids. There are truly bad things in the world but this isn't one of them!*

Case closed? Of course not. Tanker Enemy continued to use the video as "irrefutable proof" with a note that just says, "Don't listen to hoaxers who repeat that this document is false or similar lies." The video was also picked up quickly by Dane Wigington of Geoengineering Watch, who somehow overlooked the fact that it was a hoax and began to post *more* videos of similar aerodynamic contrails. In 2014, he wrote:

> *How do we know our skies are being sprayed? Because we have film footage of the crime, of jets spraying at altitude. This is the logical end of any argument or dispute on this issue. Climate engineering is not speculation, it is not theory, it is a verified fact confirmed by film footage. Those that deny what they can see with their own eyes are simply not ready to wake up. A film of the crime occurring (in this case atmospheric spraying of aerosols from jet aircraft) cannot be rationally disputed.[19]*

This was accompanied by a video of a United Airlines passenger plane leaving an aerodynamic contrail. Since aerodynamic contrails are much more sensitive to atmospheric condition than exhaust contrails, they tend to leave patchy contrails when flying through regions of uneven humidity. This is described by Chemtrail believers as being on/off spraying. But there's nothing changing with the plane, what's changing is the air that the plane is flying through.

Regions of differing humidity appear as clouds when the humidity is high enough to condense the water vapor into liquid drops. Depending on the pattern and amount of relative humidity, sometimes we see these cloud regions as isolated clouds, sometimes rows of clouds, sometimes patchy clouds, and sometimes a solid cloud layer. Thus, when the humidity is not high enough to form natural clouds and a plane flies along, then it leaves a trail only where there was already *nearly* a cloud. You get similar patterns of patchiness, but in a line.

What we've got is a normal phenomenon (an aerodynamic contrail) that is being misrepresented by people who either don't understand it, or deny it exists. Far from something that "cannot be rationally disputed," this phenomenon has a perfectly rational explanation—one that has been described and understood in books on clouds for many decades. At the very least your friend must admit that this explanation, this rational dispute, actually exists—even if they don't initially agree it is correct. The way to get them to realize the explanation is correct (and that Geoengineering Watch's flat rejection is highly unreasonable) is to show them as much about aerodynamic contrails as possible. All the books, all the websites, all the videos, all the photos, and all the scientific papers.

Aerodynamic contrails, like exhaust contrails, are real, they are a type of cloud, and they are not Chemtrails. They are not being deliberately sprayed from planes to alter the weather.

But sometimes something is.

Weather Modification

On August 24, 2017, a small Piper Comanche plane, registration N5526P, took off from San Angelo Regional Airport in Texas.[20] It flew west about eighty miles to the Pecos area where it encountered a weak storm front.

Flying along the line of clouds the pilot electronically ignited two small flares. The flares were not too different from those used by the highway patrol to create temporary traffic diversions. Cardboard tubes about an inch thick and a foot or so long, several flares are arranged in holders on each wing and wired so they could be individually ignited.

The pilot fired two more flares, then flew over to another small weather system and fired two more for a total of six flares. The next evening Hurricane Harvey made its enormous ponderous landfall in Galveston, Texas, and proceeded slowly inland to dump a historic fifty inches of rain over the Houston area—causing equally historic flooding.

Could these two events be linked? One day there's some organization conducting odd weather modification operations in Texas, and the next it's the worst rainfall in history, also in Texas? Surely these events must have been linked?

No. Texas is a big state. Pecos is over 500 miles from Houston, and at the time the flares were ignited, Hurricane Harvey was 800 miles away, out gathering energy from the warm waters of the Gulf of Mexico. There's simply no way lighting six flares in Pecos could have a measurable effect on a weather system 800 miles away. It's not clear if it really even did anything in Pecos. It would be like lighting six small flares over London, England, and having it change the weather in Rome, Italy.

And yet the conspiracy theorists still thought it was possible. They found the operations report on the Trans-Pecos Weather Modification Association's website and interpreted it as a revelation of something nefarious. The very fact that weather modification was being performed was construed by the conspiracists as validation of what they had been saying all along. Here were chemicals being sprayed out of planes. Here were "Chemtrails" modifying the weather.

The conflation between the very real process of cloud seeding and the unfounded theory of Chemtrails goes back a long way. There's a constant string of Chemtrail believers having an excited confirmatory revelation when they find out that yes, there indeed are planes spraying chemicals and modifying the weather. The conspiracy is seemingly confirmed, the debunkers are debunked, rejoice!

There are two problems with this. The first is that nobody has ever denied that cloud seeding takes place. It's something that has been done very openly and publicly since the 1950s. It has been mentioned many times in mainstream media; in fact, it even entered popular culture for a while as a topic of interest in the 1960s, to the extent that an episode of the Dick Van Dyke Show featured it as a plot device in the 1965 episode "Show of Hands," where the son needs to make a costume of a rain cloud for a part in a school play:

> **SON:** I'm a good cloud. An airplane flies over me and seeds me.
> **MAID:** Seeds you? What in the world for?
> **SON:** To make me rain on the crops.[21]

But the thing that the conspiracists seem to miss is that cloud seeding is no more "Chemtrails" than crop dusting is. Sure, it's a chemical being sprayed out of a plane, but the Chemtrail theory is quite clearly about a secret plot that involved spraying chemicals out of large planes at high altitudes, often in a clear blue sky, leaving visible trails that persist and spread, to achieve some poorly-defined nefarious end. Cloud seeding is not secret, it uses small planes at low altitudes, it's done to existing clouds, and it does not leave visible trails.

This can be a difficult thing to convey, as people are reluctant to give up the proverbial holy grail of actual-planes-spraying-chemicals. The discussion will frequently devolve into a semantic one of pointing at weather modification and saying, "Look, Chemtrails are real."

Try to stay away from semantics and focus on the knowledge gaps. Your friend probably only just heard of cloud seeding weather modification, so you should give them context. Tell them about the history of cloud seeding, explain that it has never been a secret, it uses small planes, at low altitudes, and does not leave trails in the sky.

One thing that might make your friend think that there's a connection between the trails he sees and the weather is that there actually is one. Contrails are a type of cloud, more specifically a type of high ice cloud called a cirrus cloud. Natural cirrus tends to be wispier, and often develop curls and streaks. Sometimes they form in thin layers, sometimes they are patchy. When it persists more than a few minutes a contrail will eventually start to resemble normal cirrus clouds. This was observed as long ago as 1921 in an account published in the *US Air Service Newsletter*:

> When the airplane reached a height of 26,000–27,000 feet at 11:50 a.m., a long feathery white streamer was observed forming behind a rapidly moving dark speck. The cloud was of the cirrus variety, well defined at the edges and apparently 10 to 15 times the width of the plane. The sky behind the first portion was clear blue with no clouds in the near neighborhood. . . . The whole streamer may have been 3 miles long. After 20 minutes the streamer had drifted and spread until it merged indistinguishably with the other cirrus clouds visible.[22]

For nearly one hundred years contrails have been recognized essentially as man-made cirrus clouds. One thing that has been known for even longer is that the formation of cirrus clouds can predict the weather, and in particular the existence of cirrus clouds is often an indicator of coming overcast or rainy weather. Cirrus clouds predict the arrival of warm fronts because the air of a humid warm front arrives at high altitudes first, forming cirrus clouds, then at medium altitudes forming alto clouds, then finally at low altitudes when cumulus rain clouds form. Contrails usually form a bit before cirrus clouds, so seeing contrails in a clear blue sky, shortly followed by naturally forming cirrus, is often a good clue that it's going to rain soon.

The problem here is one of false correlation. Your friend sees contrails (which he thinks are Chemtrails, because they don't fade away), and then a few hours later it gets cloudy and starts to rain. He might even notice this pattern by himself, as people have done for hundreds of years before planes ever flew. Depending on where he lives, he might not have seen any clouds for weeks, then there's these contrails, then cirrus, then rain, all in one day. You can understand how it seems suspicious.

Again, the best thing to do here is to show them old books on clouds. In particular show them a picture of a warm front moving in. Show how the picture contains the exact same things they have been seeing in the sky: high cirrus, followed by rain. It's just like that, but since the air can make cirrus it can also make contrails. That's really all there is to it.

Cloud seeding does not make clouds, it makes already existing clouds rain. Put simply, contrails are clouds that form in the same places in the sky where cirrus clouds form.

Chemtrail Planes

Another hoax that grew into a common claim of Chemtrail evidence was the one about ballast barrels, the claim that got Willie to believe:

> *The smoking gun for me, for Chemtrails, was the ballast barrels thing. When I saw that I was like, "Oh well, that proves it, oh my God." I was somewhat devastated because it confirmed that [the Chemtrails theory] was true.*

This curious bit of eternal bunk came about because of the way new types of planes are tested. All planes have what is known as a "flight envelope," which is the set of minimum and maximum conditions it is designed for—like minimum and maximum speeds and bank angles. But the flight envelope also includes the passengers and cargo. A plane must still be able to fly safely when nearly empty, and also with a full load of fuel and passengers.

These configurations need to be tested; in fact a full set of such tests is required by the FAA. Now it's not practical (or safe) to herd 400 people into a test aircraft to pretend to be passengers, so the airline manufacturers hit upon the idea of using barrels of water to replicate the weight, as explained by Boeing:

> *During flight tests in 1994, the cabin interior of the prototype 777 was filled with what at first looked like aluminum beer kegs. In fact, these 55-gallon barrels contained water. The contents of two dozen barrels in the forward cabin and a like number of barrels in the aft section were pumped back and forth to simulate shifts in center of gravity that would result from passengers moving about.*[23]

There are many similar accounts of these tests in books on aviation, and in fact they can be seen in videos of the first test flights of the Boeing 747-100[24] in 1969, and the Boeing 707 in the 1950s. Ballast barrels are a fairly common thing in the aviation industry—if perhaps unfamiliar to most people.

That unfamiliarity was taken advantage of when someone edited a photo taken by photographer William Appleton. The original photo from 2005 is of the interior of a plane that contains ballast barrels. It's in a Boeing 777-240/LR, a newly developed version of the 777. You see the barrels with tubes running between them. There's a reflective vest draped over a seat and taped to the wall there's a sign with small type fixed with yellow and black striped tape.

The doctored version is exactly the same, except the words "Sprayer 05" and "HAZMAT-INSIDE" have been added to the wall around this sign. This doctored version was then posted to various Chemtrail groups around 2007. It was very quickly debunked when the original photo was found, along with the explanation of what the barrels actually were. I wrote an article on it for Contrail Science in 2007.[25] These doctored images have been an unusually persistent bit of bunk. Despite the readily available explanations to the contrary, this photo continued to be shared as "proof" of Chemtrails, along with an increasingly large number of photos of ballast barrels in different planes.

It's this persistence (it showed up on Geoengineering Watch in 2013, six years after being explained) that paradoxically makes it a great debunking tool. It's something where you can prove conclusively that the images in question are in fact photos of ballast barrels on test planes (in a recent example from 2017, Donald Trump was photographed while being shown around such a plane). We can also demonstrate how the hoax started with the fake photo. Perhaps most importantly we can show all the people, like Geoengineering Watch, who are *still* using these photos as evidence.

A key technique in helping people out of the rabbit hole is to show them that the people they have trusted for information are not as reliable as they previously thought. You have to be careful not to simply appear to be bad-mouthing them, so if we have a clear instance like this, where they have been irrefutably wrong for years, then it will give your friend greater clarity regarding their sources of information and may allow them to ask more reasonable questions about other topics. Back to Willie:

One day, I was on Above Top Secret [a conspiracy forum], and somebody debunked the ballast barrel thing. It was like an "aha, eureka!" moment, like "wait a minute." Somebody has been lying about these ballast barrels, to make it seem like the whole Chemtrail thing is real. When the person gave me the information about the

ballast barrels, a link to the aviation site where the image was from, I looked at it
and thought: "Oh my god, the other thinking was the wrong thinking."

Chemical Tests

The ballast barrels hoax is easy to explain to people, as the idea is relatively simple to understand, and the undoctored photos are quite clear. A different piece of Chemtrail "evidence" that is frequently held up as "irrefutable" is less tractable; that is chemical analyses of soil. Not because it's empirically better evidence, but because the analysis involved chemistry, a subject that many people misunderstand. That misunderstanding has created some of the more enduring parts of the Chemtrail mythology.

The basic claim is that high levels of certain chemicals (aluminum, barium, and strontium are often listed) have been found in tests of air, water, and soil. The claim was first popularized in the pro–Chemtrail-theory 2010 documentary *What in the World Are They Spraying* (commonly referred to as WITWATS). The film was by Michael J. Murphy, but the tests used in the film were largely carried out by Geoengineering Watch. In particular a lined pond on Dane Wigington's property in Northern California was tested. The test results were shown in the documentary as:

basic
laboratory

Attention:	DANE WIGINGTON					
Project:	GENERAL TESTING					
Description:	POND SEDIMENT	**Lab** ID: 7050069-01		**Sampled:**	04/29/07	
Matrix:	Sludge			**Received:**	05/02/07	

Metals - Total

Analyte	Units	Results	MDL	RL	Method	Analyzed	Prepared
Aluminum	ug/l	375000	300	1000	EPA 6010B	05/09/07	05/05/07
Barium	"	3090	10	50	"		"
Strontium	"	345	10	50	"		"

FIGURE 8: Test results from a sample of sludge described as a "pristine" pond.

Basic Labs is a real lab and these test results are accurate. The test method listed (EPA 6010B) is a precise way of determining the amount of any element in a sample. The results from Wigington's property showed 375,000 parts per billion [ppb] of aluminum, along with some small amounts of barium and strontium. This 375,000 number has become almost mythological in the Chemtrail (or "covert geoengineering") community. It was even included in a lawsuit that Geoengineering Watch filed against the US government in 2016:

> *Per a test in Shasta County, near Redding, California, a double-lined pond (which completely prevents contact with soils), had 375,000 ppb of aluminum in the waters. Because the pond was lined, it collected precipitation; and the aluminum in the water cannot have come from the soil, but had to have fallen from the sky.*[26]

The somewhat tortuous description there is an attempt to get around a fundamental problem with all these tests: basically, that aluminum is a very common element in nature. We don't think of it as such because we think of aluminum as a metal—things like soda cans, food containers, or aluminum foil. But in nature, aluminum is a very common component of rocks, and since soil contains lots of weathered rock dust it's a common component in soil. How common? In Northern California the top few inches of the ground typically contain around 7 percent aluminum in the rock form (technically aluminosilicates). It can range anywhere from 5 to 15 percent in California,[27] but it's safe to say aluminum is everywhere.

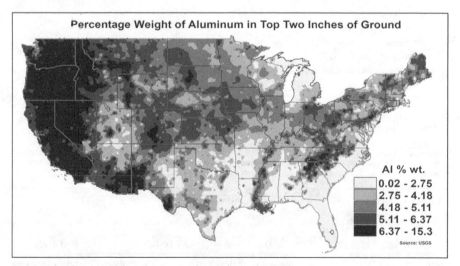

FIGURE 9: Aluminum distribution in the US topsoil, USGS.

The CDC says:

It should be noted that aluminum is a very abundant and widely distributed element and will be found in most rocks, soils, waters, air, and foods. You will always have some exposure to low levels of aluminum from eating food, drinking water, and breathing air.[28]

Since aluminum is everywhere it's going to show up everywhere, in the air, the water, and the soil. In clean water there's almost always some aluminum; it's impossible to keep out. In natural streams it's typically 100 ppb,[29] so 375,000 does actually seem really high. What's going on?

The problem here, and the key thing to explain to your friend, is that what was tested was not water, but "sludge," the stuff that was collecting at the bottom of the pond. Basically, windblown dust, or soil. It says it right there on the test, next to "Matrix." We can also see the muddy bottom of the pond in a shot in the WITWATS documentary. The pond location is also situated in a U-bend in a dirt road, and since it does not rain in the summer in California, those dirt roads kick up a lot of dust.

Since dirt is (conservatively) around 7 percent aluminum, that's literally 70,000,000 parts per billion. It does not take much dirt mixed into the water to get to 375,000 parts per billion—especially if you consider the testing procedure Geoengineering Watch posted on their site at that time (it's still there now, just not directly linked):

If you are testing a pond, then the only thing different is how you collect the sample. The very bottom of the pond is where the elements stack up. Turn your jar upside down and get the mouth to the bottom of the pond or still water . . . the older the pond the higher the readings. Turn the jar over and collect both the water and a LITTLE of the bottom sediment.[30]

That's the worst way possible of testing water. The sample will inevitably contain an entirely random amount of the dirt from the bottom of the pond, and as such will produce a random (but likely quite large) amount of aluminum. I attempted to explain this to Dane Wigington in 2013, when we had a debate:

Mick: *[Regarding] those tests, the issue basically is that sludge contains dirt and dirt is 7 percent aluminum, and so you are going to get high aluminum rate in those tests. And yet those tests were used in the film as evidence of spraying.*

Dane: *Now at face value, Mick, again, if those tests, if that material had any contact with dirt, any form of dirt, I would fully agree with you. But, this sample came from a pond that is lined with not one liner, but two, this is Firestone EPDM pond liner. It's biologically safe for fish, there is no water source into this pond except rainwater and well water. It has virtually no contact with dirt, soil or any type, kind, and that reading was high because it was taken near the bottom of the pond where there's some of the fish feces and so forth that are down at the bottom of the pond, but that was no less reassuring to us that that sort of fish sludge could contain that much aluminum, but, on that test there is absolutely no contact with the earth in any way, shape, or form.*[31]

Rather bizarrely Wigington seems to claim that a pond exposed to the air for years would be pristine. Possibly realizing how this sounds, he then says the sludge is "fish feces and so forth." But *any* pond is going to have dirt in the bottom. These types of mistaken and misrepresented tests have been repeated for many years, and many can be found in the Geoengineering Watch lawsuit, even after they were explained years ago. Another one is described as:

Pure white snow at 8,000-foot elevation on Mount Shasta had 61,100 parts per billion (ppb) of aluminum, over 4 times more than the mud beneath the snow and tens of thousands of times the expected maximum level in a snow sample. The samples also contained 83 ppb of barium and 383 ppb of strontium. The only route for these heavy metals to enter the precipitation system is from the aerosolized clouds.

The problem here is that the test result was based on a sample taken *in the middle of June.* While there is still snow on the summit of Shasta in June, it's old snow, coated with dirt. A search of photos on Flickr taken that date on Mount Shasta show images of exactly that—pockets of largely melted snow, covered in reddish brown dirt, naturally laden with aluminum.[32]

Show the original tests to your Chemtrail-believing friend, then show them the charts that show how much aluminum is in the soil. Show them the word "sludge" on the test, and the images of the muddy water in the pond, and the location on the dusty road, and the dirt on the snow. They should understand that these are simply the normal variations that come from aluminum being in dirt. Then show them that the main Chemtrail promoters are *still* using this sludge and dirty snow as evidence and ask them if they really should be treating these individuals as authorities on the subject.

Chemtrail Patents

Another very common claim about Chemtrails (or covert geoengineering) is that there are patents for it. The argument is quite simple: patents are for real things, there are patents for Chemtrails, therefore Chemtrails are a real thing.

There are three main problems with this theory that you can explain to your friend. The first one is quite simple, but something that most people don't even think of. If it's a secret government program, then why is it patented, and why are its records publicly accessible? What's the actual benefit of patenting the technology for a program that you are planning on denying the existence of for decades? Would it not be a far better strategy to keep it secret?

The second problem is that most of the patents listed are not even for technology that's related to secret spraying of climate modifying aerosols from planes. The biggest list of patents was created by Geoengineering Watch back in 2012, where they say:

> For anyone doubting the existence of the phenomenon of geoengineering/ weather modification, please take a minute to read through this extensive list of patents from America on equipment and processes used in just such programs. The evidence is clear.[33]

They then follow it up with what is indeed an extensive list of patents. The problem is that the majority of them are for entirely non-controversial technologies, like skywriting, or cloud seeding. And those are just the ones even tangentially related to "Chemtrails," there's a whole bunch more that are seemingly random, chosen perhaps because they contained a key word like "aerosol" or "spray." Here are some of the actual patents listed as "evidence":

1631753–An electric water heater

2097581–A steam generator for sterilizing lab equipment

2591988–A method for making white paint pigment

3174150–A self-focusing antenna

3300721–A radio for spaceships

3564253–A giant space mirror for reflecting sunlight onto the Earth

3899144–A towed aerial target that leaves a powder trail for visibility

3940060–A ground-based smoke ring generator for punching holes in clouds

3992628–A temporary laser shield

4347284–A snow camouflage blanket to hide tanks under

4415265–A spectrometer for analyzing chemicals

5056357–An ultrasound measuring device for liquids

5327222–A flow speed sensor

5631414–A device for measuring radiation from the ocean

6110590–A method of making silk

5486900–A device for measuring how much toner is left in a photocopier

We might forgive the compiler of the list for initially being a little over-enthusiastic, but the persistence of this list after at least five years of people pointing out the mistakes is a testament to the resistance of the list compiler to any form of correction. The last one in particular, a device for measuring how much toner is left in a photocopier, is something I have attempted to tell the Chemtrails folk about several times. I've written a detailed explanation on Metabunk[34] which I share every time it comes up, and I've even posted a comment on the Geoengineering Watch page, which was instantly deleted.

This resistance to facts is a bit disheartening, as it means this list is just going to keep coming up year after year. But it's also a useful tool. Go over this list with your friend and ask why Geoengineering Watch never changed it even after they were shown it was wrong.

If we strip away all the nonsense patents, the patents for skywriting, and the patents for ordinary cloud seeding, then there are actually a handful of patents that are genuinely for spraying stuff out of planes for the purpose of geoengineering (climate modification). This bring us to the final argument that Chemtrail promoters make regarding patents: that the existence of these patents means geoengineering is "real."

Firstly, they mean "real" in the sense of "a topic which people research." This is a nonsense argument because nobody ever denied people have been thinking about geoengineering, or even researching possible ways of doing it. What is denied is the claim that people are *actually doing it* today. And that's the second way in which patents supposedly make Chemtrails "real," the core question of their argument: Why patent something if it's not being used?

This demonstrates a fundamental misunderstanding of the patent system, missing three key facts:

- Patented technology doesn't need to work in order to be patentable
- Even if the technology works in theory, it does not have to exist
- Patents are often filed speculatively

In the United States between 1790 and 1880, when you patented something you were required to provide a miniature model explaining how your invention supposedly worked.[35] There has never been a provision in US law that an invention must meet a defined standard of operability and functionality in order to be patented. Current patent law requires only that the invention be useful for something, be a new idea, not be obvious, and be described well enough so someone could implement the invention based on the patent.[36]

There are many patents that technically fit these descriptions but are basically wild and crazy things that are either very bad ideas or could not possibly work.

One of the more famous examples of this is US 3216423, "Apparatus for facilitating the birth of a child by centrifugal force"—which is basically a circular table, to which a pregnant woman is strapped. The table is spun around until the force of the spinning pops out the baby, which is then caught in a net dangling between her legs.

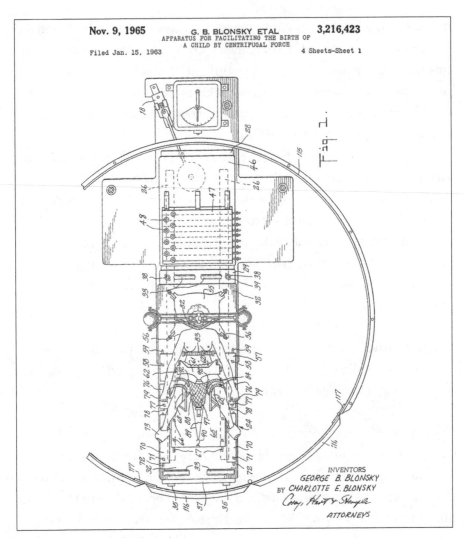

FIGURE 10: Patent for birthing device.

This idea was patented in 1963, but there is no evidence it was ever even built, and while it seems quite preposterous to most people, it seems it was patented by people who thought it might work.

An even more improbable patent is US 6960975, "Space vehicle propelled by the pressure of inflationary vacuum state," which describes a kind of flying saucer, propelled by bending the fabric of space-time. The idea is not based on any existing technology, but instead come from a very loose extrapolation of ideas in various papers on theoretical physics. It's basically no better than the description of warp drives in *Star Trek*. There are plenty of similar patents for things like anti-gravity drives, and a "full body teleportation system—a pulsed gravitational wave wormhole generator system that teleports a human being through hyperspace from one location to another."

Have these inventions been demonstrated to work? Clearly not, there are no full body teleportation devices, or warp drive flying saucers, or spinning baby extractors. Just because something has been patented does not mean it's a good idea or that it works, or even that the underlying science is correct. It just means that someone had an idea, and they wrote it down in a way the patent office would accept.

While the above patents are obviously ridiculous, or highly impractical at the least, there are some patents that actually seem sensible. Yet we still know usages of the invention don't exist because they refer to situations that do not yet exist.

The clearest examples of these are to do with manned space travel. There are many patents for interplanetary spacecraft, and for habitats for humans on other planets—and yet nobody takes this as evidence that there are people living on the Moon or flying to Mars. Here's a 1992 example of a house on the Moon which seems reasonably practical (see Figure 11).

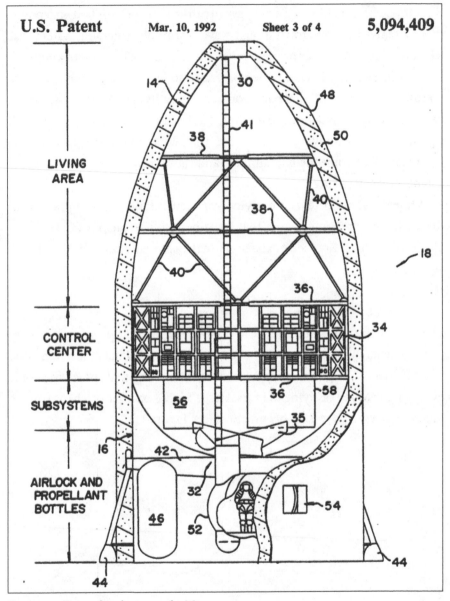

FIGURE 11: Patent for a house on the Moon.

There's also several patents for a space elevator—a stunningly ambitious method of getting things into orbit by literally building an elevator that goes up to a satellite. Like geoengineering, this is not a new idea—the space elevator as a concept was invented in 1895 by Konstantin Tsiolkovsky. Ideas about global solar geoengineering date back to around the same time, with Svante

Arrhenius proposing (in 1905) to control the levels of greenhouses gases in the atmosphere in order to create an optimal climate.

The inventions above similarly have not been demonstrated to work in the real world, but they seem much more plausible than the previous examples, and they address things that people have been seriously talking about doing in the coming decades. Like geoengineering patents, they can certainly be used as evidence that people are *thinking* about doing something in the future (people are already researching space elevators),[37] but they cannot be used as evidence that something is being done *now*.

Not only are the patent filing requirements rather vague and somewhat subjective, the current patent system has been considered by most observers to have been effectively broken for many years. Patents are awarded for the most trivial of "inventions," and hundreds of millions of dollars often change hands over ideas a child might have—like buying something with one click or allowing a user to dial a phone number by tapping it.

People recognize there is a lot of money to be made from patents. Some people file patents not because they themselves plan to develop the technology, but because they think it's possible that someone else might develop it in the future and then they can claim millions of dollars by licensing their patent. Patents just cost a few hundred dollars to file in the US (just sixty-five dollars for the provisional application).

This is called "Patent Trolling," and it's not just something that lazy individuals do. The flaws in the patent system force companies to patent every single thing they can think of, just so they will have a portfolio of patents they can counter-sue (or trade) with other companies—the 2009–2017 smartphone patent war being a good example.

These large companies file thousands of patents every year. IBM filed eighteen patents *every day* in 2012. Many of these are "just in case" patents for ideas that people in the company had. It was speculation about what they might develop in the future, not necessarily what they are developing now, and almost certainly not something they have actually finished developing.

Even if it pans out and there was some intent to use it, the vast majority of patents (95 percent) never even get used, as describe in *Wired*:

The unspoken reality is that the US patent system faces an even bigger problem: a market so constricted by high transaction costs and legal risks that it excludes the vast majority of small and mid-sized businesses and prevents literally 95 percent of all patented discoveries from ever being put to use to create new products and services, new jobs, and new economic growth.[38]

Yes, there are geoengineering patents, and patents for houses on the Moon, and Mars spaceships. But it does not mean that geoengineering is happening today any more than it means there are men on Mars.

All a patent means is that the person or company who applied for that patent thought that they could either make money from the patent in the future, or they should add it to their patent portfolio, just in case they wanted to develop the technology in the future, or possibly use the patent to make a deal later.

Even if your friend ignores all of the above, even if they continue to insist that patents are evidence of something, there's still the fundamental objection to the whole "patents as evidence" theory—if you wanted to do something in secret, then why would you let private companies patent evidence of that thing?

Bottom line here? *Patenting a thing does not mean that thing exists, or even that it works.*

One With, One Without

Once you've explained to your friend that contrails can actually persist, and that modern planes are actually more capable than old planes of making contrails, then a subtler version of the theory emerges. The idea now is that planes are simply making contrails when they should not be.

There's two versions of this claim. In the first you are shown (or told about) two planes that are in the same general area of the sky at the same time. One of the planes is leaving a long thick persistent trail. The other plane is not leaving a trail or is just leaving a short trail that quickly dissipates.

The solution to this puzzle is incredibly simple. One plane is *flying higher than the other.* It's simply impossible to tell how high one plane is relative to another when both planes are over thirty thousand feet and the planes are of different types.

A useful way of demonstrating this is to use a Flight Tracking app like FlightRadar24. Use the app to find the altitude of the planes and show that the planes they thought they were at the same altitude were actually at different altitudes, often over a thousand feet apart.

Then note that contrails only form in altitudes where it's cold enough and humid enough. If one plane was at that altitude (and leaving a contrail), and another was just above or below that altitude, then it might not leave a contrail.

The second version of this claim involves an attempt to determine the weather at the altitude a particular plane is flying and noting that the humidity is too low for contrails to persist. This would be a valid approach, however the measure that is commonly used is perhaps the least accurate for this purpose—data from weather balloons.

Weather balloons are a bundle of instruments (thermometers and humidity sensors) attached to a balloon. A balloon is released and rises through the atmosphere. As it climbs it radios back the readings from these instruments until the balloon finally bursts and falls back to the ground. These balloons provide a nice snapshot of a vertical slice of the atmosphere. Unfortunately, balloons are released only twice a day, and from stations that are 200 to 400 miles apart. Atmospheric conditions can rapidly change over just a few minutes, and in just a few hundred *feet*, so these widely spaced snapshots are generally useless for finding the conditions at any one particular spot in the sky. Compounding the problem is the fact that some types of humidity sensors on the weather balloons actually stop working at around the same temperatures that contrails form.[39]

Some far better sources of atmospheric humidity data are the various weather forecasting models which combine the weather balloon data with continuously updated fine grained data from ground stations, aircraft, and NOAA satellites. This produces a much higher resolution chart of what the humidity is at any location and altitude at any given time. Invariably the appearance of contrails matches up on these charts with the presence of appropriately high humidity (above 60 percent). There are multiple independent outlets who distribute this weather information.[40] Show them to your friend and have them check the local humidity against their observations and contrast it with the measurements the Chemtrail promoters make using only weather balloons.

Government Admissions

There are two ways in which people claim the government has "admitted" to Chemtrails or covert geoengineering. The first is to point at weather modification (discussed earlier). In this case you've just got to explain to your friend what weather modification actually is: cloud seeding to make it rain or snow more, something that has been openly done for sixty-plus years.

The second way is to point to people in government or academia discussing possible future geoengineering, and then claiming that's an "admission" of current geoengineering. Here's an example.

> Chemtrails have long been regarded as "just another wacky conspiracy theory," but what's your excuse when a former CIA Director [John Brennan] himself admits that the government is spraying our skies? . . . Indeed, whereas the notion of secretive government programs spraying chemicals into the sky is often deemed a conspiracy, the government seems to be openly engaging in essentially the same practice now.[41]

The best approach here is to first get your friend to actually read the full context of what Brennan said. Then get them to look at the actual current state of geoengineering research (discussed earlier). Here's what Brennan actually said:

> Another example is the array of technologies—often referred to collectively as geoengineering—that potentially could help reverse the warming effects of global climate change. One that has gained my personal attention is stratospheric aerosol injection, or SAI, a method of seeding the stratosphere with particles that can help reflect the Sun's heat, in much the same way that volcanic eruptions do. . . .
>
> As promising as it may be, moving forward on SAI would raise a number of challenges for our government and for the international community. On

the technical side, greenhouse gas emission reductions would still have to accompany SAI to address other climate change effects, such as ocean acidification, because SAI alone would not remove greenhouse gases from the atmosphere.[42]

Notice Brennan's use of the future tense: moving forward "would" raise challenges, SAI "could" help reverse the effects of global warming. It's simultaneously speculative and uncertain—a description that reflects the current state of geoengineering research and explains why nobody has actually done it yet. The government has not admitted Chemtrails or admitted geoengineering. They have simply admitted something that nobody has ever denied—that geoengineering is something we might consider doing in the future.

Chemtrails has been described as a gateway conspiracy theory. It's easy to get into as you can just look up to the sky and see the "evidence." There's a large number of seductive and seemingly authoritative videos on the topic and a laundry list of different claims of evidence to suck people in. But it can also be a gateway *out of* the rabbit hole. As we've seen in this chapter, all the claims of evidence have been addressed, often in great depth. If someone takes an honest and unbiased look at the evidence then it's very unlikely they will continue to believe. The challenge is in getting them to take that first look. In the next chapter we'll meet Stephanie, who resisted looking at contrary evidence for a long time, until she was helped by a friend.

Chemtrails Summary and Resources

Key Points to Convey to Your Friend
- Contrails can persist and spread, and have done so since the 1920s.
- There's a lot more air traffic now than thirty years ago.
- Modern jet engines make more contrails, not fewer.
- Trails that come from the wing are aerodynamic contrails.
- Weather modification has been done openly since the 1950s and does not leave trails.
- Photos of barrels on planes are ballast barrels on prototype commercial planes.
- Aluminum is everywhere, so it shows up in all tests of air, soil, and water.
- There are genuine concerns about geoengineering, but we've not even got to the stage of doing tests yet.
- Patents don't mean something works, and why would you patent a top-secret program?
- Different planes leave different length contrails at different altitudes.
- The government has not admitted to a secret geoengineering program.

Additional Resources
- Contrail Science—**contrailscience.com**
- Metabunk's Chemtrail forum—**metabunk.org/Chemtrails.f9/**
- Atmospheric Optics—explains many of the things people see in the sky like halos and Sun dogs that are sometimes associated with "Chemtrails." **atoptics.co.uk**
- NASA's Worldview—satellite images of clouds lets you see contrails and weather in a global context. **Worldview.earthdata.nasa.gov**
- FlightRadar24—Lets you identify planes leaving contrails so you can see they are mostly normal commercial jets. **Flightradar24.com**
- Earth Wind Map—lets you see the complexity of wind and humidity patterns aloft. **earth.nullschool.net**

CHAPTER EIGHT

Stephanie – A Former Chemtrailer

Stephanie Wittschier was once a firm believer in the Chemtrail theory. She was deep down the conspiracy theory rabbit hole for many years but was eventually helped out of it by a friend. Stephanie now runs the German website *Die Lockere Schraube* (The Loose Screw) where she uses her experience to help others.[1] Her tale is a validation and inspiration for skeptics and for the friends of people seemingly trapped down the rabbit hole.

I asked her what it was that got her interested in the Chemtrail theory, and as usual it started with video.

The video I watched that got me into Chemtrails was a [2011] interview with Brigitta Zuber on Alpenparlament.tv [a German New-Age alternative media site].[2] Brigitta Zuber talks about Chemtrails, which can be seen for much longer than "normal" contrails. "Contrails disappear after a few seconds," Zuber said. After seeing the interview, I looked out of the window and I saw contrails which didn't disappear after a few seconds. They stayed there for several minutes. At that time I didn't know very much about meteorology and so I thought that she could be correct and that there was something wrong.

Looking at the sky and seeing persistent contrails after watching a video about them is a surprisingly common thing. People don't pay much attention to the sky other than looking to see if rain is likely, or if the Sun is shining. The persistent contrails have always been there, but people don't pay them any attention until they are pointed out. If they trust the source that told them that contrails should not persist, then they believe what they are seeing is unusual, despite the fact that even in the conspiracy version it's supposedly been going on since the 1990s and they never noticed until they watched a video about it.

Stephanie was very much predisposed to believe what Zuber was saying:

At that time, I had a tendency toward conspiracy theories. For example: I believed in the existence of aliens and that they visited us in the past and will visit us again. Also, I believed that spaceships and dead aliens were kept in Area 51 and that the government lies to us about that. When I was eleven years old, I was given a book about aliens and that stuff. Since then I have believed in that. Later I also believed in things like "Bigfoot" or "Nessie" and that the pyramids in Egypt were maybe built by aliens and that Stonehenge is also kind of mysterious. But that's it. I didn't notice other conspiracy theories. They didn't really interest me.

Then, I think it was at the end of 2009 or at the beginning of 2010 (I can't remember when it starts exactly) I saw a documentary about 9/11 together with my husband. There were some strange "facts" given and so I was curious, and I did some research on the internet about 9/11. Looking back, it was the dumbest thing I could have done. Because I discovered more and more about conspiracy theories. All of these theories have little connections and so automatically, especially on the internet, you acquire the knowledge of all these other theories.

Then I went to web pages about conspiracy theories, Kopp Verlag [a German publisher of New Age books], Alles Schall und Rauch [Sound and Smoke, the German equivalent of "Smoke and Mirrors"—a general conspiracy site], Alpen Parlament. Then, very quickly after that, I signed up to some online forums, which are about those theories. Of course, there was "Allmystery.de," the biggest forum about conspiracy theories in Germany.

She got sucked down the rabbit hole, deep down. Just like me, Stephanie had a childhood interest in things like UFOs, monsters, and aliens. I think the fact that she found herself down the rabbit hole, and I did not, speaks to the sometimes-random nature of who gets sucked in and who steers clear. It can boil down simply to happening to watch a particular video or read a particular book.

Stephanie was convinced, she ventured down the hole, and it began to affect her life.

When my friends started to question me about this I think I was angry with them. I can remember a controversy with one of my friends. Then she unfriended me, which I can understand now. With my husband and my closest family members it was another kind of story, luckily. They didn't search for conflict and they let me do my own thing, although they didn't believe in those theories. If they didn't do it this way, maybe I would have lost them all. That's what happens to many of the conspiracy theorists. They lose their friends and families because of their beliefs.

And it turns out that *not* losing her friends and family was the key to her escape.

There was a certain turning point in my belief in conspiracy theories. I had a best friend in the conspiracy scene. We believed in the same things, had the same opinion and we were in the same forums too. But then, suddenly, she turned her opinion. At least I thought that it was sudden. In fact, she had done research for months and started to question all these theories.

She started to ask me and other conspiracy theorist questions that we had heard over and over again from the skeptics. She posted links to web pages from "paid disinformation agents" (in German we would say: "bezahlte Desinformanten"). I couldn't understand why she did that. We exchanged some PMs and I asked her why she behaves like a skeptic now. I was very disappointed and then we had a serious conflict because my disappointment turned into anger. I thought she was some kind of traitor. We broke up completely and I didn't get over that for months.

Because of that, I started to have a look at the skeptics' side myself and eventually started to question the theories as well. You know, for conspiracy theorists it seems like you change your opinion in just one day. In fact, it is a process that took several months for me. The result was, that my friend was right. All of these theories I believed in were more than idiotic. I was shocked. You know, it's a tough situation when you realize, that you made a goof out of yourself all these months. I really felt ashamed for my behavior. Today I couldn't believe why I was so dumb.

It was not simply her friend's prompting that led to this. There were also some events within her group that led her to be somewhat more open to their ideas. Much like Steve in Los Angeles, seeing her peers embrace things that seemed clearly on the other side of her personal demarcation line led Stephanie to question if she'd drawn the line in the right place.

Regarding the belief in Chemtrails specifically: When I read in a Chemtrail-group on Facebook, that they wanted to blind pilots with a laser pointer I was shocked. I thought: "You can't do that. What about the other people on that plane? Why do they want to end so many lives?" That was the beginning of my freeing from the Chemtrail scene. Then I heard about a project. Someone in the group wanted to look at what was really in the Chemtrails. He wanted to give 200 Euros [to a project investigating if Chemtrails were real] and asked in our group who else wants to be a part of that project. Me and a few other users wanted to be a part of it. But other users were not very pleased, including an admin of that group. The result was that the thread was deleted and the users who wanted to be a part of the project were thrown out of the group. That was the second time, that I thought different about the Chemtrail scene. So, I started to look at the skeptic side and I started the research.

My friend told me at that time, that I should go to pilot forums and talk to pilots in person. After the two incidents I thought: "Ok, let's do it." So, I talked to the users there, I talked to meteorologists and to the skeptics I first hated so much. Bit by bit I discovered that the Chemtrail scene is more than idiotic and that I was wrong. And I discovered how dangerous that scene really is. E.g. Laser-Pointer-Attacks or MMS-enemas with children (to get out the "Chemtrail poison"). As I became fully aware of the situation I decided to "fight" against the Chemtrails scene and other conspiracy scenes. It needs to be observed, so that other people cannot be lost to it.

To get out is tough. When the other conspiracy theorists know about that, they start a smear campaign against you. From insults to death threats, anything can happen. Even today I receive death threats. You'll be very happy that none of them live nearby or they could visit you, or something else. It's a serious threat!

Conspiracy theorists think that they question everything. In fact, they just quote other theorists and mostly they receive their "knowledge" from YouTube videos. So did I, at that time.

I'm pretty sure that many of the "extreme" conspiracy theorists I know actually got to the point where they discover that they believed in nonsense. But they didn't go the way I did. They stopped questioning things, turned around and went back to their beloved scene. They can't admit that they behave like goofs and so they return to their feel-good zone. To be skeptical seems too hard for them I guess.

I did more research, like I said, otherwise I could have been in a psychiatric ward today. But I found the emergency brake. I did admit in public that I made mistakes. I apologized to the skeptics and to my former best friend for my behavior. We put our differences aside and became friends again.

I asked Stephanie about the language of conspiracy theories and debunkers.

*We say: "**die rote Pille nehmen**," "take the red pill," which is similar to "going down the rabbit hole." Some conspiracy theorists think that The Matrix movie is not fictional, it's some sort of documentation and it shows how things really are. Another phrase I use is "im Verschwörungswahn" or "Verschwörungssumpf sein." It means that you are so deep in the conspiracy scene (or the conspiracy swamp), that you believe everything and that you are unable to see that all you believe in is just idiotic. The way out of it seems impossible and you don't listen to anything skeptical anymore.*

Besides the help from her friend, Stephanie found guidance in other places.

Sites like "Contrail Science" or "Metabunk" did help me to get out of the "conspiracy madness" and to get more knowledge about reality. The skeptics on AllMystery .de did help me very much through my thinking process. They were very patient and finally I really understood their answers and wanted to hear them.

This is quite heartening for skeptics like me who spend time explaining things online. It often seems like a Sisyphean task, as the explanations just bounce off the conspiracist's wall of "evidence." But what Stephanie's case illustrates so well is that it takes time, and every little bit has some effect. You just can't fully see that effect until the person emerges into the light. We should expect that our seemingly logical and fact-based points will initially be rejected, the important thing is to keep the discussion going.

Another interesting thing is the role of conspiracy forums. While there's obviously some great debunking going on in the skeptics forums like InternationalSkeptics.com and Metabunk.org, there are also a large number of skeptics who like to post in the lion's den. In the US there is AboveTopSecret.com. AllMystery.de is essentially the German equivalent. Much valuable debunking occurs directly on these sites, and people's lives have been changed for the better because of it.

CHAPTER NINE

9/11 Controlled Demolitions

All 9/11 conspiracy theories can be summed up with one phrase: "9/11 was an inside job," the implication being that it was not some outside adversary (al Qaeda) that committed the attacks that brought down the World Trade Center, badly damaged the Pentagon, and crashed a fourth plane into a field. It was instead some elements operating within the upper reaches of power, people inside the US government and the military–industrial–financial complex that is actually running the country.

This simple "inside job" descriptor covers a wide range of very different theories. Of any conspiracy theory, 9/11 has perhaps the broadest and most detailed spectrum of variations. At the bottom end of the spectrum there are the most trivial forms of "let it happen," where the Bush administration simply ignored some warning signs of some kind of attack. In this lightest version the Bush administration is guilty of little more than negligence. It's hard to call it a conspiracy at all.

Slightly further along the conspiracy spectrum is the idea that the attacks were not planned by al Qaeda, but instead by various other actors. Typically, these will include elements within Saudi Arabia, then Israel, and (as we go further up the spectrum) the United States. These theories are based largely on circumstantial evidence and it's rare for people to be deeply invested in them. Generally, you'll find these stay in the realm of "just asking questions" and "where there's smoke there's fire."

At the far end of the spectrum are people who think that the attacks never actually happened, that what we saw on TV that day were computer generated animations. They think there were no planes and that the towers were destroyed with explosives, possibly nukes, or space energy beams. The evidence for these theories is generally highly specious and depends on deep misunderstandings of physics and a significant disconnect from reality. There are many rebuttals to these theories even within the core 9/11 community because the average 9/11 Truther does not want *their* theory to be associated with the "nonsense" on the other side of their demarcation line.

But squarely in the middle is the most common form of the "9/11 was an inside job" conspiracy, and the one you are most likely to encounter with your friend: it's the idea that there was some kind of *controlled demolition*. In this theory, the towers were not brought down simply by being hit by planes and the subsequent fires, they were supposedly pre-rigged with explosives, all the way from top to the bottom, and these explosives were carefully detonated in sequence, in an attempt to make it look like a more natural collapse.

The root reasoning behind the controlled demolition theory is quite simple: the collapses of the towers just didn't look right. Indeed, if you look at footage of the two collapses of the Twin Towers of the World Trade Center (WTC1 and WTC2), they do look very strange. It seems almost like one continuous explosion making its way down the tower. There's what looks like huge plumes of material being ejected out sideways, and then this wave of "explosions" just comes straight down, the building disintegrating and exploding before your eyes.

There are numerous objections and rebuttals to the controlled demolition theory, some more technical than others. But the bottom line is that the extraordinary events of 9/11 were simply something none of us had ever seen before, so it's easy to rely on our limited personal experience and say, "buildings just don't do that!" The challenge in helping your friend out of the 9/11 rabbit hole is largely composed of explaining to them exactly *how* buildings can do that.

The first challenge you are likely to encounter on this road is one of symmetry of perception between you and your friend. You think you are right. You think you understand that planes flew into the buildings, badly damaging them, and that the buildings burned. You understand that steel was weakened by the fire, that columns failed, and that the structure below could not stop the falling mass as it gathered speed. Perhaps you know these things when you start out trying to explain them to your friend, or perhaps they will come up later. Regardless, at some point you are going to think you have it figured out, and what remains is to explain it to your friend whose mind has simply been brainwashed into these strange beliefs.

Unfortunately, your friend has *exactly* the same idea about you. They have done their research. They have watched many videos, and they have read many web pages, perhaps even read several books on the subject. They are deeply convinced they have the science on their side. They feel that the controlled demolition conclusion is both obvious and irrefutable because they know that jet fuel can't melt steel beams, and they can see explosions going off in the videos, and they can see the concrete being pulverized and multi-ton sections of the building being thrown sideways by explosions. They think that *you* have been brainwashed, and their challenge is to explain things *to you*. They think *you* have a mental block in seeing the obvious truth, and it's *your* terror of the implications of being wrong that keep you pushing what seems to them to be a blatantly false narrative.

Here's a Facebook comment made about someone discussing 9/11:

The guy knows the truth he just needs the attention. Or he doesn't know the truth in which case he has a learning disability and I wouldn't want to make fun of him. Some of us have worked with concrete, jack hammered concrete, or just lived enough for common sense to work. Best guess is this guy just needs the attention and knows better. A lot of people died on 9/11 and it takes a pretty sick mind to want to argue those lies the way he is.

Is this discussing a 9/11 conspiracy theorist? No, that quote was actually about *me*, and was written by someone who thinks that it's blindingly obvious that the World Trade Center tower collapses were controlled demolitions. Despite the fact that I'm reasonably intelligent, with a good grasp of basic physics, people constantly tell me I'm stupid when I question a particular piece of 9/11 evidence. This isn't limited to 9/11 either—one of my TV appearances to discuss Chemtrails only came about because the Chemtrail believer they first interviewed suggested the hosts interview me. He later told me that this was because I was so stupid, he hoped I would come across badly.

It's very important that you realize the depth of your friend's belief, the degree to which they reject criticism, and that you try to figure out what his perception of you is. What does he think motivates you? Does he think you've just been misled, or that you are stupid, or crazy? Some in the 9/11 Truth community go as far as to suggest a psychological basis for your rejection of their theory, like a mental illness. In the 2012 video "9/11 Explosive Evidence—Experts Speak Out," a number of supposed experts are interviewed about the usual things like explosives and physics, but they also interviewed psychologists who had found their way down the rabbit hole (nobody is immune). Psychologist Fran Sure said:

> What is common to every [person who rejects 9/11 controlled demolition] is the emotion of fear. People are afraid of being ostracized, they're afraid of being alienated, they're afraid of being shunned. They're afraid of feeling helpless and vulnerable, and they're afraid that they won't be able to handle the feelings that are coming up. They're afraid of their lives being inconvenienced . . . of being confused . . . [and] of psychological deterioration. They're afraid of feeling helpless and vulnerable.[1]

Also, from licensed clinical psychologist Robert Hopper, PhD:

> 9/11 Truth challenges some of our most fundamental beliefs about our government and about our country. When beliefs are challenged or when two beliefs are inconsistent, cognitive dissonance is created. 9/11 Truth challenges [our] beliefs that our country protects and keeps us safe and that America is the 'good guy.' When this happens, fear and anxiety are created. In response, our psychological defenses kick in [to] protect us from these emotions. Denial, which is probably the most primitive psychological defense, is the one most likely to kick in when our beliefs are challenged.

Similar sentiments were expressed on Facebook,[2] in rather more colorful language:

> Just an aside. I don't speculate as to what motivates any individual skeptard. No point in calling them shills without definitive proof. I suspect skeptardism, also

known as sciencetardism, is a form of brain-damage, possibly related to autism, that makes skeptards cling to authority and the seeming comfort of simple explanations for the actual, fucked-up, confused mess that defines the reality we live in. Thus, assumptions like "conspiracy theories are impossible" and "scientific consensus as an incorruptible principle." Like a religion, skeptards fit their pre-conceived conclusions into the box of that set of basic assumptions about reality. It's known as scientism—the belief that "all truth is derived through science." Via all kinds of bluff and bluster and convoluted reasoning, they pimp and proselytize. Skeptards are incapable of original thought. They thrive in the seeming safety of a hive-mind. That's why they all sound exactly the same. A mental dysfunction of right-brain cripples. It's a tough world when all you can do is focus on trivial details, but can't connect the dots to see the big picture. After all, that's 'conspiwacy theowy.'

This belief in the "mental dysfunction" of skeptics arises from a serious problem for people in the 9/11 Truth movement: why don't more people believe them? From their perspective the conspiracy is crystal clear. They think they have the evidence, and indeed that the evidence is so blatant, so obvious, that the only way someone intelligent could possibly examine it and then deny it would be if they were either in on the conspiracy or if they have some kind of mental block, some pathological subconscious terror of upsetting their worldview.

This is what your friend will almost certainly end up thinking of you at some stage in the process. At first, they will simply assume you don't know what the evidence is, it's just not been explained to you, or you've not done your research. But if you persist in trying to help them out of their rabbit hole, then at some point they are either going to label you as a shill or brainwashed and incapable of facing the truth.

This problem exists in all conspiracy theories, but it's particularly strong in 9/11 theories. A big part of the reason why is that they feel that they have the authority of science on their side, and this itself is largely due to an organization called Architects and Engineers for 9/11 Truth.

Architects and Engineers, Experts?

Architects and Engineers for 9/11 Truth (AE911 Truth) was founded in 2006 by a practicing architect, Richard Gage. The goal of the organization as stated on their website is to "Build a better world with sound scientific research into the destruction of the World Trade Center."[3] The year of foundation (2006) comes after NIST's 2005 report on the collapse of the Twin Towers, but well before the 2008 report on Building 7 (a common piece of evidence promoted by conspiracists). An early (2007) version of their website presents what is inadvertently perhaps the most accurate summation of their core problem.

> The 6 years since 9/11/01 has given us the time and space to emerge from the hypnotic trance of the shocks of these attacks and to rationally evaluate the existing and new evidence that has become available.
>
> Architects and Engineers are trained to design buildings that function well and withstand potentially destructive forces. However, the 3 high-rise buildings at the World Trade Center which "collapsed" on 9/11 (the Twin Towers plus WTC Building #7) presented us with a body of evidence (i.e. controlled demolition) that was clearly outside the scope of our training and experience.[4]

AE911 Truth is very important to most people who think that there was a controlled demolition of the three World Trade Center buildings, and it's important precisely for the reasons that Richard Gage identified—the physics of the impacts, the fires, and the collapses themselves were all events of a type and a scale that were unprecedented in our experience. Very few people have the scientific background to understand the physics of the events, so they turn to others to help them understand it. While Gage starts out acknowledging this, he quickly notes in the same paragraph that now he thinks the task is *not* outside the scope of his training and experience:

> There is however a growing body of very solid evidence regarding these "collapses" that has emerged in the last couple of years—gaining ground even in the mainstream media. This new evidence casts grave doubt upon the theories of the 9/11 building collapse "experts" as well as the official reports by the 9/11 Commission, FEMA, and NIST.

It lays out a solid convincing case which architects & engineers will readily see: that the 3 WTC high-rise buildings were destroyed by both classic and novel forms of controlled demolition. You will find the evidence here in our website as well as at the linked websites. We hope you will find the courage and take the necessary time to review each section thoroughly.

Where was AE911 Truth getting this evidence? The first three set of linked websites back in 2007 (911Truth.org, WTC7.net and 911research.wtc7.net) were created by Jim Hoffman—a key figure in the history of the 9/11 Truth movement, but *not* exactly someone with training and experience in the field of controlled demolition. Hoffman, a software engineer, has spent many years theorizing about 9/11. In one article he proposes, *in all seriousness*, that the Twin Towers' high speed progressive collapses were facilitated by a thin layer of explosives hidden inside 1.8 million ceiling tiles, combined with a layer of explosives literally painted onto the core columns and other explosives hidden inside fire extinguishers.

The new ceiling tiles with embedded thin-film explosives and wireless detona-tors are installed throughout every other floor of the Tower. In all, each Tower gets 500,000 of the large tiles and 400,000 of the small tiles.

With workers swapping in new tiles at an average rate of two tiles per minute per worker, it takes a team of forty workers 187 hours to retrofit an entire Tower. The work is performed in three weeks and weekends of night shifts, emptying one truckload per night, with the truck parking inconspicuously in the WTC subterranean parking garage. . . .

Once descent of the Tower's upper block begins, the thin-film explosives [in the ceiling tiles] are detonated via the wireless control system just ahead of the descending wave of destruction below the crash zone, as well as in the upper portion of the descending block.[5]

This illustrates two things. Firstly, the incredible fixation that Truthers have on the idea that the collapses were impossible from fire and gravity alone. So strong is this belief that eventually they come to think of the towers as nearly solid blocks that need to be blown up with explosives on every floor. A huge part of debunking 9/11 conspiracy theories is simply getting past this idea. In Hoffman's case, the belief is so strong that it requires literally millions of individual explosives for the observed collapse to seem plausible in his mind.

Secondly, the fact that Hoffman is one of the primary sources of evidence for AE911 Truth shows the flaw in their appeal to authority. Your friend will likely bring this group up with a question like, "How can so many scientists think that 9/11 was controlled demolition?" The implication is that we are supposed to trust AE911 Truth because they are experts, and there are a lot of them.

Yet the foundational evidence of AE911 Truth came not from structural engineers or demolition experts with skyscraper experience. It came largely from people like Jim Hoffman, or Steven Jones (a retired physicist), or even David Ray Griffin (a theologian). If we look at the position statements of the people who make up AE911 Truth we do not find sophisticated analyses; instead we find people who had initial suspicions, who read the websites and watched the videos, and for the most part are no different to ordinary people who have been sucked down the rabbit hole. Here are a few representative statements from the AE911 Truth petition signatures page.

"I have known from day-one that the buildings were imploded and that they could not and would not have collapsed from the damage caused by the airplanes that ran into them."

"I suspect the whole 911 thing as being badly misrepresented by authorities, a cunning deception by social deviants; we must seek truth in everything, otherwise our culture and all educational institutions will be transformed into endoctrination [sic] centers. Pray for this nation."

"What architecture student hasn't watched a video of demolitions in structures class? It seems obvious that the first two buildings were destroyed by demolition. The disconnect happened when the media failed to explore the possibility. When I learned about WTC-7, I knew for certain."

"Buildings collapsed all by controlled demolition methods. Fire and impact were insignificant in all three buildings. Impossible for the three to collapse at free-fall speed. Laws of physics were not suspended on 9/11, unless proven otherwise."

"The buildings fell at or very near free fall speed and into their own foot prints! The second airplane went through the corner of the building, with most of the fuel burning outside the building. Furthermore, it did not hit the structural core. Yet it fell the same as the first tower!!! Impossible!!"

"There is little doubt that the collapses were caused by controlled demolitions and the aircraft impacts were causal ruses."[6]

The root fallacy that is in play here is the fallacy of the *argument from personal incredulity.* In almost every case the signers are basing their belief in controlled demolition on the fact the collapses looked odd, and they could not figure out how such odd-looking collapses were possible without explosives. Far from bringing their expertise to bear on the problem, they have instead failed to use that expertise, and instead relied firstly on their initial gut assessment and visual perception of the extraordinary events, and secondly on confirmation of the assessment from compelling YouTube videos.

Even when the source of AE911 Truth's information seems like a genuine expert, there's a pretty good chance they have made some mistakes along the way if they are buying into a mistaken conclusion. Steven Jones, for example, is an actual scientist, a physicist. He claimed in 2006 to have found the chemical signature of thermite in a sample of residue on a 9/11 memorial sculpture that was created with some of the site debris. This claim was repeated as fact by AE911 Truth for the entire lifetime of their organization. Then in 2018, twelve years later, I looked into one of his claims and in the course of a day discovered that Jones had been reading the graph wrong, seeing a peak for manganese where there was none, and the sample was probably just slag from the cutting of the beam with an oxyacetylene torch.[7] The elements he claimed to have identified were in fact textbook examples of common misidentification for that specific technique.

Richard Gage, the head of AE911 Truth, used that incorrect analysis in his presentations for ten years. He also used a photo of a column from the World Trade Center that had been cut at an angle. Gage claimed it was cut with thermite charges during the collapse. I was able to find a series of photos of the column, intact and cut, that were both taken *six weeks* after the collapse.[8] Gage, a licensed architect, had been using it incorrectly for ten years. Steven Jones, the originator of the "nanothermite" hypothesis, has also continued to use it. Expertise does not mean infallibility, especially when making extraordinary claims.

The challenge for you with your friend is to *show* them this fallibility in AE911 Truth, this tendency to make mistakes. It's a tricky thing because the same people who think that all the other scientists are incompetent and/or corrupt often also hold AE911 Truth in the highest regard. Avoid simply telling them that AE911 Truth don't have any credibility, and instead focus on *showing* your friend where AE911 Truth has gotten it wrong, and where they sometimes make the most ridiculous claims. Tell your friend about AE911 Truth's history, and how they did not come to "9/11 Truth" via any special investigation. They originally came to it via the work of non-expert conspiracy promoters, like Hoffman or Griffin. And show them where their nominal experts, like Jones and Gage, made mistakes that went unnoticed for over a decade.

Molten Steel and Nanothermite

A foundational mistake made by AE911 Truth and others in the 9/11 Truth movement is the idea that molten steel was found in the debris pile weeks after the collapse. Here's David Chandler writing on AE911 Truth's site:

> *There were pools of molten steel under the rubble piles of Buildings 1 and 2 (the North and South Towers) and Building 7 that remained molten for weeks after the building collapses, indicating a continuing energy source.*[9]

He's referring to a common belief held by 9/11 Truthers that there were literal pools of white-hot molten steel underneath the rubble pile. They don't just mean red-hot steel girders, they mean *liquid steel*. They mean molten metal bubbling and flowing like lava for weeks after the collapse.

Steel has a very high melting point, 2500°F, far higher than the temperatures reached in building fires, and a lot higher than the temperatures that the steel columns of the building reached (as calculated by NIST). So pools of molten steel would be very suspicious.

The first problem here is that there was no physical evidence that these pools ever existed. There were certainly some eyewitness accounts of molten *something* being seen at various points under the rubble pile, but it could quite easily have been aluminum, lead, or glass that simply melted at some point in the underground fires. But leaving that aside, if there were "pools" of molten steel then at some point those pools would have cooled and solidified into huge chunks of solid steel. No such chunks were ever found.

(One large piece of debris known as "the meteorite," on display at the 9/11 Memorial Museum, is sometimes presented as such a chunk. But it's just compressed never-melted steel, concrete, and other debris. It contains pieces of paper, so obviously was not part of a pool of molten steel which have incinerated any paper.)

Chandler says "indicating a continuing energy source" because no matter how hot a supposed pool of molten steel could be, it is physically impossible for it to remain fully molten for more than a few hours, due to the radiation and conduction of heat.[10] To get around this, Chandler introduces an equally impossible notion—that something was continually burning under the rubble to keep the steel molten. Exactly how this something managed to burn for weeks without being consumed, and also managed to be right next to the pool of steel, is an implausible conundrum.

That special "something" is, in 9/11 mythology, "nanothermite."

Thermite is an incendiary substance, typically a mixture of aluminum powder and iron oxide power. It burns at an incredibly high temperature, 4000°F, making it capable of melting through steel. One of the byproducts of the combustion is molten iron. If you were to burn a ton of thermite you'd get half a ton of molten iron as well as molten steel from whatever you were cutting. So if you are looking for something that made "pools of molten steel," thermite is a good candidate.

The origin of the nanothermite myth can be traced back to before June 22, 2003, in an article by Michael Rivera of our friend Willie's then-favorite news source, What Really Happened:

> *The collapse of the "spire" is consistent with a thermite reaction pooling molten iron into the central area of the WTC basement and subsequently melting the core columns, thus inducing the collapse which occurred seconds later.[11]*

The early version of the thermite myth makes no real sense as melting the core columns (of the World Trade Center Towers) would require a literal lake of molten iron. Filling up the 200 feet square basement level of either World Trade Center tower to a depth of one foot with molten steel would take 40,000 cubic feet of steel. This lake would eventually have solidified into a huge slab of iron weighing five thousand tons, which, again, was not found during cleanup. But this wild speculation is the origin of the nanothermite story. Rivera was actually something of a thermite fan having suggested it in previous conspiracy theories like the 1996 plane crash that killed Ron Brown[12] and possibly in the explosion of TWA Flight 800.

At this point in time Truthers thought it was just regular thermite. Later writers upgraded it to "super thermite," and then settled on "nanothermite." Nanothermite sounds exotic, but it's really just regular thermite with finer particles, so it burns faster, somewhat more like a conventional explosive than regular thermite. This results in a lot *less* melting of steel. It's also not something that has ever been documented as being used for demolition either before 9/11, or in the seventeen years after.

How would burning thermite keep steel molten for weeks? It couldn't. One of the largest thermite burns in history was performed by *Mythbusters*, who used half a *ton* of thermite to cut a car in half.[13] This burned out in just a few minutes, and the area quickly cooled enough to approach shortly after that. Even if you had an implausible giant block of hundreds of tons of thermite, it would still be consumed in a few hours at most—and much less if it were the faster-burning nanothermite.

AE911 Truth talked themselves into a corner with their embrace of the "pools of molten metal" evidence. Even though the pools did not exist (and physically could not exist), they had to concoct an explanation for them consistent with their controlled demolition theory. Once they had fixed on thermite they went looking for more evidence to confirm their theory.

In the dust of the building they found two things: microscopic iron spheres and small chips that were red on one side and gray on the other. They claimed these were both evidence of thermite. But the iron microspheres were actually something expected in the dust from the fires, as explained in 2012 by R. J. Lee, an independent analyst who extensively studied the dust.

What about the iron microspheres? The iron has a thin layer of rust flakes that can easily be removed by sticky tape. The iron is heated red hot or hotter and subjected to hurricane force blast furnace-like wind. The iron flakes are liberated as small particles and some iron is vaporized. Like drops of water, the iron flakes form molten spheres that solidify and the fume also condenses into spheres, the most efficient geometrical form. . . . The formation of iron and other type spheres at temperatures obtainable by the combustion of petroleum or coal based fuels is not a new or unique process.[14]

But burning flaking iron is not the only way of making iron microspheres. I've done some experiments and discovered numerous other methods, as have scores of scientists and researchers.[15] You can easily make millions of microspheres with common construction activities such as arc welding, angle grinding, or hammering steel. Shielded from the elements inside the building, these microspheres could last for decades, until the eventual collapse of the building freed them into the dust. Microspheres are also created by mundane items like the sparks from cigarette lighters or flint ignitors used by steel workers. The pre- and post-collapse fires would create them by burning small pieces of iron dust (you can make then yourself by burning steel wool) but also by burning other things, like laser printer toner, and printed documents (the toner contains iron). After the collapse there was a lot of iron that needed to be cut to be removed, and the cutting produced huge amounts of iron microspheres which ended up in the general site dust.

Not only are iron microspheres an expected component of the dust, that same dust was also *missing* an expected product of thermite combustion. When thermite burns it makes twice as much aluminum oxide (by volume) as it does iron. When we burn thermite we find many white aluminum oxide spheres and a large number of hybrid "eyeball" spheres with a dark iron sphere embedded in a larger white oxide sphere. These were absent from descriptions of the dust. The simplest explanation is that the iron spheres are from mundane sources, and thermite was not used.

What of the curious red/gray chips? They look exactly like paint chips. I've found some of these chips myself simply by hitting a red painted steel wheelbarrow with a hammer.[16] Identical looking paint chips flaked off. They were red on one side, had a gray layer of rust on the back, and were attracted to a magnet. When I heated them in a flame they formed microscopic shiny spheres of some iron-rich material (it was magnetic). This is all *exactly* like the flakes that AE911 Truth claims were leftover thermite. All the steel used in the World Trade Center was painted with red primer paint like my wheelbarrow. Therefore you would expect to find millions of tiny chips of this primer paint in the dust.

AE911 Truth claims that analysis shows it's not paint. But their own tests do not really back up such claims. They determined the gray layer to be iron oxide, which is rusted iron—exactly the same as the chips from my wheelbarrow. The red side was more complex, but was consistent both chemically and morphologically (shape and texture) with paint ingredients like kaolin, containing aluminum and silicon. They also found several different types of chips with different chemical signatures and behaviors—consistent with various different types of paint.

How do you convey this to your friend? It can be difficult as things like chemical analysis are not easily digestible to most people. You need to identify what simple information they are missing. They will generally have been aware of this thermite claim as a simple list of evidence: molten steel was found, and the signature of thermite was found in the dust in the form of iron microspheres and red/gray chips. What you need to convey is: solidified pools of previously molten steel were never found during cleanup, iron microspheres are an expected result of a raging fire in a steel frame building, and the red/gray chips look and act just like the expected paint chips from the millions of square feet of red painted steel.

With iron microspheres, get a lighter and make some of your own. Just spark the lighter a few times over a sheet of white paper over a magnet, and you will get a dusting of microspheres. Look at them through a microscope. Bring up how AE911 Truth likes to quote expert R. J. Lee about how many iron microspheres were found in the dust (sampled months after the collapse in an area with lots of steel cutting activity), and yet they ignore or reject his explanation of those spheres. This cherry picking of an expert's testimony is not new for AE911 Truth; they also did it with Danny Jowenko.

Danny Jowenko

On the home page of Architects and Engineers for 9/11 Truth there's a list of the top eight pieces of evidence for controlled demolition of Building 7. Number seven on that list is:

Expert corroboration from the top European controlled demolition professional.[17]

That sounds very compelling—if the top European demolition professional agrees it was controlled demolition, then that's it, right? The expert in question is Danny Jowenko, who died in a car accident in 2011. Back in 2006, he was interviewed for a Dutch TV show *Zembla*.[18] They showed him footage of Building 7 collapsing in the *Loose Change* conspiracy documentary. He had not seen this before and his immediate assessment was that it looked like a controlled demolition.[19]

What most 9/11 Truthers don't know is that Zembla also interviewed him regarding the collapses of the Twin Towers, where he had nearly the *opposite* assessment. He thought that controlled demolition of the towers was basically an impossible idea. Here are some excerpts of that section, which is almost always omitted or evaded by Truthers:

Jowenko on why the second building hit was first to fall:

You clearly see that the building that was hit first was hit higher, so it went last because there was less weight to bring it down. That's essential knowledge for anyone who knows anything about demolition: you have to use the building's own weight.

Jowenko on why explosives were not used:

You'd place the explosives below, of course . . . Yes, that's how you get the full weight. That's a [gift]. The less you have to blow up. But the tower collapsed top down. It collapsed at the exact location where the plane hit and heated it. . . . It can't have been explosives, as there was a huge fire. If there had been explosives, they would already have been burned. What's more, before being burned their igniters would have gone off at 320 degrees Celsius, so they'd have detonated sooner.

Jowenko on why it looks like there are explosions in the sides of the building:

You also see, as it were, the bolts springing loose at each turn. It had a very strong core, and the beams were pretty long, but they're joined, and it was 410 meters tall. The energy is very uneven. So, every vertical column has to carry a certain weight at a slightly different moment from its neighbor, so to speak. It can't bear it, so it breaks to pieces across its entire length, bolts and all. It comes loose, all the way down. And the side structures, also strong because of the wind stress, which is how the building was built, were mainly pressed outward. . . . [As the firefighter in the video says], it simply gave out. At every level the weight was too much.

Jowenko on the plausibility of rigging the building with explosives:

Don't tell me they put explosives on all 100 floors. That's not possible. . . . It would take a year [to place all those explosives] and prepare them and hook them up with all the cables down there.

If you accept this expert's opinion on WTC7, why not accept his opinion on WTC1 and 2? This can be a very powerful argument to raise with your friend—largely because they have never heard of this before. They quite possibly have been showing Jowenko's opinion on Building 7 to everyone but were totally unaware of his analysis of the other towers. Show then the conflicting opinions, and they have to pick one, or dismiss both, or start to think about things in a bit more depth. Dissonance sows the seed for escaping. Then ask them why AE911 Truth has this on their list of their best evidence. What does that say about their other "best evidence"?

The Plane That Hit the Pentagon

On September 11, 2001, American Airlines flight AA77 flew into the side of the Pentagon. It was 9:37 a.m. on a clear Tuesday morning. The Pentagon is one of the world's largest office buildings where over twenty-five thousand people work. It's surrounded by huge parking lots, then encircled by several major freeways packed with commuter traffic. Beyond the freeways there are tall office buildings, malls, apartment buildings, and hotels, all within two miles. Tens of thousands of people had a clear line of sight to the Pentagon, and hundreds of them saw the plane fly by, over Interstate 395, and into the side of the building.

And yet there are still many people who think that AA77 did not hit the Pentagon, and that, despite the impossibility of pulling it off without anyone noticing, the Pentagon was actually hit by a missile.

The Pentagon plane debate is a divisive and important issue in the world of 9/11 conspiracy theories. For many people it's where they draw the line. They think it's plausible (even obvious) that explosives brought down Building 7, but the idea that a missile hit the Pentagon without anyone noticing falls on the other side of the line in the "silly theories and disinformation" category.

If your friend also draws the line here, there's two interesting things you can discuss. Firstly, you can ask why they don't believe. What exactly is it that makes a missile hitting the Pentagon a less realistic scenario than sneaking tons of explosives into buildings and setting it off with millisecond precision after arranging for planes to fly into the buildings and having them burn fiercely for an hour, all without leaving any physical evidence? Is there anything about their reasoning around the Pentagon that might also apply to the World Trade Center?

Secondly, you can discuss why so many people believe in this Pentagon no-plane theory, or at least give it credence. As the more extreme theory you'd think it would be significantly less popular than the controlled demolition theory. But they are actually very close, with the line not needing to move very much. A 2006 Scripps Howard poll found the exact same percentage of Americans (6 percent) thought that it very likely that "The Pentagon was struck by a military cruise missile in 2001 rather than by an airliner captured by terrorists" as those who thought it very likely that "The collapse of the Twin Towers in New York was aided by explosives secretly planted in the two buildings."[20]

The Pentagon no-plane contingent seems to encompass the majority of 9/11 "controlled demolition" Truthers. These people include founding luminaries of the 9/11 Truth movement,[21] such as David Ray Griffin (by far the most popular 9/11 conspiracy author), Kelly David[22] (Chief Operating Officer at Architects and Engineers for 9/11 Truth), Craig McKee (AE911Truth contributor and podcaster), and Niels Harrit (9/11 conspiracy scientist who popularized the idea that nanothermite was found in the rubble).

If your friend thinks that flight AA77 hit the Pentagon, then ask him why they think that all these "experts" got it wrong. Could it be that trust in their expertise was misplaced? Could it be that even architects and engineers, and scientists of all stripes, can also be sucked down the rabbit hole?

But if your friend thinks that without a doubt flight AA77 did not hit the Pentagon, and that instead it was a cruise missile, then you are going to have to explain why that is incorrect. The details are too complex to get into fully, but very briefly what you have to show to your friend is:

The eyewitness testimony: Read *all* the eyewitness reports. There's strong and consistent accounts of an American Airlines jet hitting the Pentagon. These accounts match across dozens of independent people. Don't just read them, watch and listen to the accounts of eyewitnesses given to the press on the day of the attack.[23]

The context of the eyewitnesses: Show your friend where the Pentagon is, next to the freeways. Show them Google Street View images of the Pentagon from the freeway. Show them that if it was not a plane, or if the plane flew away, then hundreds of people must have seen it.

The actual size of the hole: They frequently will have seen photos of either the small hole in an inner ring of the Pentagon where some piece of debris punched through, or the notch above the entry hole formed by the tail. Show them the actual lower hole on the ground floor,[24] which matches the size of the plane.

The aircraft debris: There were lots of pieces of the aircraft scattered around both inside and outside the Pentagon—some visible in photos taken minutes after the attack. There were also lamp posts that were knocked over, one hitting a car. Was this all planted *without anyone noticing*?

The damage from the wings: To the left and the right of the bottom hole are columns that were damaged by the wings. You can see the damage in photos taken immediately after the impact, and in later photos.[25]

The relative ease of the maneuvers: The flight into the Pentagon was a descending 300° turn, then a straight-in shallow descent into the south wall.

This sounds hard (and conspiracy theorists will tell you it was) but is actually a standard maneuver for reducing height, simply called a "descending turn." All you have to do is reduce power a bit, turn the yoke to the right, and wait until the turn is done. It's taught to pilots with only a few hours of experience.[26]

The ASCE report: *The Pentagon Building Performance Report*[27] published by the American Society of Civil Engineers details exactly what happened in great detail, showing you on a column by column basis what was damaged by the plane's fuselage, engines, and wings.

9/11 Truther reports: The more science-based 9/11 Truthers draw the line just this side of the Pentagon no-plane theory and have devoted substantial effort to explaining why it is false. Show your friend the paper "The Pentagon Event"[28] by seven members of Scholars for 9/11 Truth (and Justice). In it they provide a detailed explanation of the evidence mentioned above. And these are people who think the World Trade Center buildings were destroyed by controlled demolition.

There are a lot more, but these are the core aspects of the false belief. Given time and care, discussing these with your friend can help move their line a bit closer to reality. When that happens (and it has settled in) then you can ask them why they trust AE911 Truth's expertise when their COO and many of their members think that no plane hit the Pentagon.

The Missing $2.3 Trillion

One reason suggested for why the conspiracy had to include a precise missile or bomb attack on the Pentagon is to cover up $2.3 trillion of missing Pentagon money. Here's David Ray Griffin:

> *The official trajectory [of AA77 into the Pentagon] was impossible to execute . . . so why would Pentagon officials choose to explode bombs in that part of the Pentagon and then claim it was struck by a hijacked airliner? . . . One suggested answer puts together two facts: First, the day before 9/11, Secretary of Defense Rumsfeld stated at a press conference that the Pentagon was missing 2.3 trillion dollars. Secondly, one of the most damaged areas was the Army's financial management/audit area.*[29]

Much like the Pentagon missile issue, the $2.3 trillion issue is a wonderful opportunity to bring your friend some perspective. Nearly everyone in the 9/11 Truth movement (and quite a few outside it) think that $2.3 trillion was missing from the Pentagon. It was never missing; it never even really existed. Rumsfeld settled on the $2.3 trillion figure by adding up all the times transactions were recorded in the Pentagon's accounting systems where transactions did not come up to code for accounting standards.

The first thing you've got to do is give your friend some perspective on large numbers. $2.3 trillion is a very large number, in fact it's larger than the entire US budget spending for that year. It's also *ten times* as large as the entire national defense budget ($274 billion in 1999, the accounting year in question). It's literally impossible that the Pentagon could have lost ten times as much money as they actually spent.

How did they get to this number? Consider what would happen if you wrote down all the dollar amounts on your pay slip, bank deposit statements, W2 tax statement, transfer statements from checking to savings, ATM withdrawals, new car loan, checkbook record, credit card receipts, actual receipts, bills, mortgage statements, mortgage payment records, mortgage tax statement, credit card statements, credit card payments, monthly bank statements, receipt for a sale of an old car, deposit slip for sale of old car, title transfer for old car, payment schedule for new car, and your tax return. Maybe you make $60,000 a year, but this all adds up to $600,000 or more! Now let's say you keep some of these records in a shoe box under the bed and lose the rest. You lose track of records totaling $300,000. Does that mean you've lost $300,000? Obviously not. Same thing with the Pentagon.

This was explained by the Department of Defense in April 2001:

> *For the FY 1999 financial statements, the auditors concluded that $2.3 trillion transactions of the $7.6 trillion entries to the financial statements were "unsupported." DoD notes that many of these entries included end-of- period estimates for such items as military pension actuarial liabilities and contingent liabilities, and manual entries for such items as contract accounts payable and property and equipment values. DoD would further note that the "unsupported" entries are "not necessarily improper" and that documentation does exist in many cases, albeit, not adequate for the auditing standards imposed.*[30]

There are very real problems with excessive and wasteful spending by the US military. There continue to be significant problems with accountability. However, these problems are being spun as money being "missing" when it is not. This was repeated by RT (Russia Today) in 2013, who promoted the false idea that $8.5 trillion was missing,[31] when in fact $8.5 trillion was the entire US military spending over a period of sixteen years (1996 to 2012). The story that they spun was again about the accounting standards for the money that was spent, not about tangible money that went missing. RT has continued to push versions of this false story almost every year. The most recent instance was in December 2017; this time the amount in question ballooned to $21 trillion.[32]

This can be a difficult concept to convey to your friend, but it's *very* worth giving it time. It's a relatively simple error once you understand what is going on, and yet one that has been remarkably persistent at all levels of 9/11 Truth and the conspiracy world in general. Once your friend has moved this particular claim to the other side of their demarcation line, you can ask them what other things people might have gotten wrong. Ask them if their source made a genuine mistake, or are they misleading them? Given that so many people claim there actually is $2.3 trillion missing, despite it being explained many years ago, maybe they should exercise a little more caution about the other "facts" those people have been promoting for years.

Useful Information

Most people who subscribe to the 9/11 controlled demolition theories have a rough idea of what the evidence is. There's lists on websites like AE911 Truth, but their acceptance of those lists of evidence is based on two things. Firstly, it's based on the authority and expertise of AE911 Truth, which as we've seen does not always hold up to scrutiny. Secondly, it's based on an incomplete understanding of what the items on the list actually mean. This incomplete understanding comes from a lack of information, and so to help your friend better assess those lists, you need to supply them with useful information and the necessary context to understand it.

A detailed explanation of the claimed evidence, the useful missing information, and the real explanation would comprise another book in itself. Here

I'm just going to list a number of items and sources of information. There are also numerous other online sources of 9/11 debunking information available if you want more details.

The NIST Reports. The often denigrated NIST reports are an incredibly detailed account of the investigations into the collapses of the World Trade Center buildings.[33] There's quite a few of them, but if you just want to look at one to start, then I recommend "NCSTAR 1-9, Structural Fire Response and Probable Collapse Sequence of World Trade Center Building 7."[34] It's long, but try to get your friend to just scan through it and get a sense of the detail and the amount of work that went into it.

The NIST FAQs. The FAQs are relatively short lists of questions that the general public and 9/11 Truthers frequently raise about the NIST investigations. There's one for the Towers,[35] and one for WTC7.[36] If your friend has questions, then look them up here first. There's a surprising amount of information in them that most Truthers are unaware of.

Vérinage. Demolition without explosives involves dropping the top part of a building on the lower part to destroy the entire thing.[37] It's similar to how the WTC Towers collapsed. If your friend is convinced only explosives can bring a building down then they might be unaware of vérinage. They should also look at it for "squibs" (expulsions of air, which Truthers claim are explosives) and "pyroclastic-like clouds" (expanding dust clouds which Truthers say must be from mid-air explosions).

Controlled Demolition with Explosives. These are VERY LOUD! Listen to actual controlled demolitions.[38] They are a series of super loud bangs which would echo through the city and easily be captured on the audio portion of video tapes from half a mile away. Compare to the WTC collapse video tapes which have no such bangs.[39]

Rock Falls. When a large object like a rock, a concrete floor, or a building full of drywall falls and hits other objects, then firstly a lot of dust is produced, and secondly the dust billows upwards in cauliflower shaped clouds. AE911 Truth says this is suspiciously like a "pyroclastic flow" from a volcano. But look at video of giant rock falls in Yosemite (with no explosives) and you see the same effect.[40] It's just crushed rock (or concrete and drywall in the case of the WTC). The clouds form because of the displacement of air by a large falling object. It was hot in places because of the large fires.

Weakened Steel. When steel gets hot in a regular fire it weakens. Two great examples of this are the heating of an iron bar by Purgatory Iron-works,[41] and the heating of a structural beam with jet fuel by *National Geographic*.[42] In both cases the steel failed, just like it did in the World Trade Center. It's a fun experiment to try yourself if you have a blowtorch.

Black Smoke. A common misunderstanding is that black smoke indicates "incomplete combustion, low temperature fires." But if you look at fires of jet fuel *in an open field*,[43] you see there's lots of black smoke, but there's obviously plenty of oxygen available, and the fire still heats the steel beam to over 1100°C, until it fails.

Slender Column Buckling. Tall columns need to be braced to maximize strength. In the World Trade Center this was done with floors. If the floors failed then the column strength was greatly reduced. Show your friend some video examples of slender column buckling with and without bracing.[44] This was a significant factor in all the WTC collapses.

Static versus Dynamic Forces. A common misconception is that since the lower part of the building could support the upper part normally, then it should be able to stop it once it was falling. This misunderstanding is based on not knowing that a moving, dynamic load generates vastly more force than a non-moving, static load. Even a very short drop of a few inches will multiply the force many times. Show them some illustrations of static versus dynamic forces.[45]

The WTC Debris Pile. What's interesting about the pile is the total lack of any evidence of controlled demolition. If you look at any failed connection in any photo, you see that it's basically been ripped or snapped apart. The core columns do not have cut ends, they just snapped apart at the shallow welds after the floors were stripped away. There's a lot of photos of the pile at various stages. Look at the columns. How did they fail?

Debunking Websites. Your friend probably spends a lot of time looking at videos and websites on one side of the argument. Try to get them to look at a broader range. Say you'll read their website if they have a look at something like 911Myths.

Plasco. Since a mantra of the 9/11 Truth movement is that "no steel framed building has ever collapsed from fire alone," then when the Plasco building collapsed from fire in Iran, killing twenty firefighters, AE911 Truth were forced to either walk back that particular piece of evidence, or also become Plasco Truthers. They choose to become Plasco Truthers, bizarrely adopting an obscure old commercial building in Iran as the fourth building to be covertly destroyed with secret nanothermite.

9/11 Summary and Resources

Key Points to Convey to Your Friend

- You've done your research, and are willing to listen to them.
- Architects and Engineers for 9/11 Truth seem largely no different from ordinary 9/11 Truthers, they are just architects and engineers who got sucked down the rabbit hole.
- Pools of solidified molten steel were not found in the debris, and physically could not have remained molten for weeks. Nor could nanothermite burn for weeks.
- They should at least read the NIST FAQs.
- AE911 Truth cherry-pick their expert testimony. R. J. Lee found microspheres, but also explained them. Jowenko said WTC7 was controlled demolition, but he insisted the Towers were not.
- A plane hit the Pentagon, but lots of Truthers think it did not.
- $2.3 trillion was not missing, but lots of Truthers think it was.
- When shown an example of a steel framed building collapsing from fire, AE911 Truth simply said that was nanothermite too.
- Most pieces of Truther evidence have answers you can google. At least consider the alternate explanation before accepting just one side.

Additional Resources

- Metabunk's 9/11 Forum—wide-ranging discussions.
  metabunk.org/9-11.f28/
- International Skeptics 9/11 Forum—Even more wide-ranging.
  internationalskeptics.com/forums/forumdisplay.php?f=64
- 9/11 Myths—comprehensive coverage of the main myths. **911myths.com**
- 9/11 Guide—lots of interesting information. **sites.google.com**
  /site/911guide/
- Debunking 9/11—Older site but still useful. **debunking911.com**
- NIST's WTC Disaster Study—all the reports, FAQS, and videos. **wtc.nist.gov**

CHAPTER TEN

Karl – Temporary Truther

Karl is a comedian and YouTube content creator who has made some excellent videos explaining how World Trade Center Building 7 fell. He posts under the name "Edward Current." In the comments on his first debunking video[1] (which I *highly* recommend you watch and try to show to your friend) he briefly mentions that he used to be a "deluded 9/11 Truther." I got in touch and asked him how he got interested in the Truth movement.

> *It was about 2007, 2008. I don't tell people how long it lasted because it's kind of ridiculous, but maybe a week. I learned about Building 7 and I watched what they usually show, which is the very last global collapse without the penthouse, and I thought "oh my god!" Then I read this page that was like "Nine Things That Don't Make Sense." And then I was kind of all-in, I even drafted a letter to* Frontline, *the PBS show, which I love, it's an amazing show. I never sent the letter.*
>
> *Then about a week into it I asked my brother about it and he just kind of shrugged and was like "Occam's Razor," the least complicated explanation is probably right and that would be fire.*
>
> *Then I was like, "Why did it go straight down?" and he just shrugged and said, "Gravity points straight down," and I was like, "That kind of makes sense."*

This was initially surprising to me, as I had never heard of a 9/11 Truther who got in and out so fast. His route in was quite a common one—in that people see a video, they can't explain the video, then they start watching other things and it starts to make sense. But what happened with Karl after that was different. He was lucky to have a brother that he trusted. His brother essentially caught him just before he fell deep down the rabbit hole. Karl was simply not in the hole long enough to get stuck.

It's possible I'd get sucked in. I like to think I'm one of those people who is happy to be shown I'm wrong, and then change my mind. They all say that, but it's kind of like a Dunning–Kruger thing. While I wouldn't say I'm the most skeptical person I think a lot of Truthers would say, "I'm really skeptical and I don't want to believe the government did this to us, this is an objective viewpoint." But I like to think I'm a bit more introspective about my own flaws that might lead my thinking astray.

Dunning–Kruger is a common mental bias that we all share where we tend to overestimate our abilities.[2] The lower our abilities, the more we overestimate. In nearly every field 90 percent of people in it rate their ability as "above average." During his very brief stint as a Truther, Karl did what most over-confident Truthers do: he did his own research.

I was just looking at videos and reading documents and things like that. But of course I only did it on the one side, the side I was looking for at that moment. I didn't look at the other side of the story. I wanted to believe this so I was drawn to all those materials. But then you get the "echo chamber" effect.

He's seen this in the more long-term Truthers he's been interacting with:

I think the more emotionally invested you are in it, the less likely you are to be able to pull yourself out, just because you've spent so much time and energy on it. Some people just do not like to be shown wrong. They react with anger, they resist it. But then other people are just like, "I'm a dumb shit."

Another thing that Karl had in his favor was that almost nobody around him was really into conspiracy theories.

The only one is this very smart guy, who listens to KPFA, which is big on 9/11 conspiracy theories. He saw one of my [debunking] videos and I got an email from him just out of the blue saying, "How can you believe this stuff!" He was just laying into me about the guys living in caves [a reference to Osama bin Laden] with box cutters. You know, all of the language. And we had a couple of exchanges and then just agreed to disagree. But other than that, no, I don't really know anybody who latches onto these things.

Now Karl spends some of his free time making debunking videos which he puts on YouTube, and then more time in the YouTube comments section responding to feedback regarding his videos. We discussed his motivation.

The thing that keeps me engaged is checking back and looking at comments. Seeing the kind of intellectual dishonesty that's going on. I just want to try to show people that their thinking is flawed. Part of it is I'm a guy and guys tend to be warriors, and they want to show people that they're right. It's funny that there's so few women that do this kind of thing. It's always men kind of at war with each other.

It's also kind of an intellectual exercise. I really enjoy finding logical fallacies and things like that, making analogies, finding parallels between things, etc. It's psychologically fascinating, and the physics of the engineering is fascinating.

As Edward Current, Karl also put a little time into hoaxing 9/11 Truthers. He created a fake video that appeared to show a close-up of Building 7. Unlike the other videos that had been made public, this one showed windows being blown out and the sounds of explosions. The 9/11 community initially lapped it up, getting very excited over this new "100 percent proof" of controlled demolition. But people quickly noticed a few problems with the video. For one if you flipped it and slowed it down you found it was actually just the real-world footage, with some added explosions. An extra giveaway was the UFO, a small flying saucer that zoomed across the shot as the building collapsed. This had mixed reactions.

[With my videos] it seems like people are divided right down the line. Some people, when you tell then they've been fooled they're like, "I'm a dumb shit, ha ha ha, you got me. Next time I'll be more careful." But the other half just get pissed off and lash out and go running.

That kind of divided the community, and to this day there's some people who think I'm a liar, I didn't hoax that video and that it's a real video, and they challenge me to put up the source files. Even though I put in the flying saucer and hidden messages and things. I made it so ridiculous, I actually made it fall faster than free fall. It doesn't match any other video that people had seen [but they still thought it was real].

Prevention is good. My goal is not necessarily to convert Truthers back to non-Truthers. But there are people who are just starting to get into it. If they can see a reasonable explanation maybe it will prevent them from going down that path.

Ultimately Karl was only very briefly down the rabbit hole, about a week—his descent stopped by his brother. He came out of it as anti-conspiracy as some of the people who had been down in years before they found their way out. It's almost as if a trip down the rabbit hole acts to inoculate you against getting sucked in again. Once you recognize your own cognitive failings and set out to change them, then there can be no going back. Exposure to conspiracy theories can suck you in, but once you've seen the (real) light, you see the rabbit hole for what it really is, and you are no longer tempted to remain.

His brother was able to stop Karl's descent relatively easily because Karl had not spent much time reinforcing his belief, compared to a case like Stephanie's. But the strategies and tactics that worked were basically the same. A friend gave him some information that he was missing. The same techniques for helping people seem to apply regardless of how long they have spent mired in the conspiracy swamp. Karl's brother gave him some simple answers in a non-judgmental manner. Stephanie's friend gave her some simple answers. Karl was quicker to accept the new perspective than Stephanie, but both of them escaped with the help of their friends.

CHAPTER ELEVEN

False Flags

Shortly after ten p.m. on October 1, 2017, sixty-four-year-old Stephen Paddock smashed two windows of his hotel suite on the thirty-second floor of the Mandalay Bay Hotel and opened fire on the country music festival crowd on the other side of the Las Vegas strip. Alternating between the two windows, he shot for several minutes using different weapons from the twenty-three he'd brought into the suite over the preceding three days. Fifty-eight people were killed and nearly 500 injured. It was the worst mass shooting by a single individual in the history of the US.

The massacre unleashed a nearly unprecedented wave of conspiracy speculation. People seemed unwilling to accept that Paddock had acted alone. Raising echoes of the infamous grassy knoll claims surrounding JFK's assassination, there were accounts of a "fourth floor shooter." Just like with the police audiotapes from the JFK assassination, people began to analyze the audio of the event to try to prove there were multiple gunmen. People who had never been to Vegas asked how Paddock could have smuggled the guns into his room past security (in suitcases, they don't have metal detectors in hotels). They asked how he could possibly have got so many guns (legally, purchased from gun shops in Nevada). They asked why someone would do such a thing (we may never know, but we know that some people are insane and/or evil).

Beyond these more-or-less reasonable questions, some people even asked if the event had really happened at all. Were people just pretending to be shot? Why were there so few photos of dead bodies? Why were some people laughing after the event? Why did some people run, but others just stood there? The questions were ridiculous, specious, and endless, but they all were based on a specific suspicion (or sometimes belief): that the Las Vegas shooting was a "false flag."

A false flag is an event that is intended to draw blame to a third party for an ulterior motive, often as a way of indirectly attacking that third party. One might use a false flag as a pretext for invading a country by first staging an attack using your own resources, but outfitted to look like your enemies' forces. For example, in 1953, under "Operation Ajax" the CIA helped arrange a series of attacks in Iran by hiring infiltrators to pose as members of the communist Tudeh Party. The goal was to promote fears of a communist revolution to create conditions for the removal of the democratically elected government.

The idea of a false flag has a long history in conspiracy culture. The attack on Pearl Harbor that precipitated America's entry into the Second World War, while a genuine attack, has long been something that people suspected that the US government had some foreknowledge of. The "false flag" aspect was the supposedly manufactured perception that it was a "sneak attack," which created a higher level of outrage in the American public, leading to stronger support for the war and greater compliance with conscription.

Similarly, there are low-level conspiracy theories around 9/11, the "LIHOP" (Let It Happen On Purpose) scenarios where people anxious for "a new Pearl Harbor" have, by inaction, allowed the attacks upon the US. The 9/11 conspiracy theories further along the spectrum (the MIHOP theories) are also largely "false flag" theories. In both cases the supposed intent being to give the US a pretext for the "War on Terror," restricting civil liberties, and allowing for more action in the Gulf region to further the goal of Big Oil and the Military–Industrial Complex.

The phrase itself was not particularly common in conspiracy culture in the first decade of the twenty-first century.

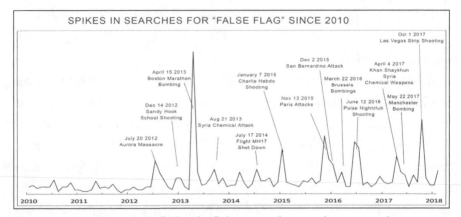

FIGURE 12: Google Trends for "False Flag," showing spikes at each mass casualty event.

"False Flag" only started to be used reflexively by theorists after a sequence of three events: The Aurora theater shooting in July 2012, the Sandy Hook school shooting in December 2012, and then the largest spike after Boston Marathon Bombing of April 2013. Then there were minor spikes (in Google search traffic) for a chemical weapons attack in Syria and then the shooting down of flight MH17 over the Ukraine. Two attacks in 2015 in France (*Charlie Hebdo* and the Paris attacks) got the Alex Jones treatment, followed by the San Bernardino shooting and the Pulse Nightclub shooting in Florida.

Other than the Boston Marathon Bombing, these spikes were all eclipsed by the conspiracist frenzy following the Las Vegas shooting, where it quickly became apparent it was going to be in a class of its own as a target for conspiracy theory false flag speculation. Even though a similar number of people died in the Pulse nightclub shooting, the reaction seemed very different.

This is possibly due to the unfamiliarity of the perpetrator. The public had become used to mass killers being psychopathic young men or, more recently, Jihadist terrorists. A young Muslim extremist killing forty-nine in a gay nightclub might possibly be a false flag, but it's also not without a context that explains it. An old white man acting alone with no discernible motive just seems inexplicable, and the conspiracy mind, abhorring the inexplicable, rushes to fill the void with an explanation: a false flag, designed to take away our guns.

When you discuss with your friend why they believe in the false flag theories—regardless of if they are 9/11 related, LIHOP or MIHOP, or even fake mass shooting "take away our guns" scenarios—one question that should be given serious consideration is, "Is a false flag operation on US soil something that the government would even do?"

At a very simple level, the answer seems to be yes—in the mind of the conspiracist the government is a corrupt, evil entity who would stop at nothing to reach their goals of world domination. Suggestions otherwise will evoke responses like "don't tell me you trust the government" and "they have done it before."

The former can be partly addressed by explaining exactly how much you trust the government (probably, like me, not a great deal). The "they have done it before" claim deserves looking into. When exactly have "they" done this before? What exactly did they do? How did it work that time? What considerations went into the decision to perform a false flag in the US? What are the risks? What are the rewards? Was it worth it?

By the far the most common example that comes up when you ask your friend for evidence that the government does this type of thing is Operation Northwoods.

Operation Northwoods

Operation Northwoods wasn't really an operation, it was a document. A 1962 report, a few pages long, suggesting a number of possible "false flag" operations that might be carried out in order to justify a US invasion of Cuba. It stated:

> *The desired resultant from the execution of this plan would be to place the United States in the apparent position of suffering defensible grievances from a rash and irresponsible government of Cuba and to develop an international image of a Cuban threat to peace in the Western Hemisphere.*[1]

The report is often presented as evidence that the US government plans false flag operations, and hence is used as supporting evidence that the 9/11 attacks were "an inside job." In more extreme interpretations, it's used as evidence that the 9/11 attacks were actually fake. Operation Northwoods exists at the messy intersection between truth and fantasy.

We don't have much to go on at all—there's the fifteen-page Operation Northwood's document titled "Justification for Military Intervention in Cuba," which consists of a few memos, and a list of nine ideas.

1. Harassing the Cubans in Cuba until they pushed back, giving a "legitimate provocation as the basis of military intervention."
2. Stage fake attacks in or near the Guantanamo Bay US base.
3. Fake a ship being blown up by the Cubans, followed by a fake rescue operation, and fake casualties.
4. "Develop a Communist Cuban terror campaign in Miami" and other US cities.
5. Stage intrusions into the airspace of the Dominican Republic, burn some Dominican sugar cane fields with Russian-made incendiary devices.
6. Use a fake MIG aircraft to buzz a real US civilian airline flight, creating eyewitness accounts.
7. "Hijacking attempts against civil air and surface craft should appear to continue as harassing measures condoned by the government of Cuba."
8. Fake the appearance of Cuba shooting down a civilian aircraft, with fake casualties.
9. Fake the appearance of Cuba shooting down a military aircraft, with fake casualties.

But there's very little recorded about the proposal beyond this memo. That does not stop people writing about it as if it's some kind of well documented event in our history. Nobody has ever given interviews about their involvement, most of the high-ranking people listed in the document are now dead. The documents came to light in May 2001 in the book *Body of Secrets* by James Bamford.[2] Note this is just before the 9/11 attacks of 2001, which might explain why some quickly made the connection.

The conspiracy community had a variety of interpretations of the documents.

JESSE VENTURA: *The military [and the Joint Chiefs] wanted to use our CIA and military to attack certain parts of the United States and make it look like Castro did it, so that the country would get up in arms and support an invasion of Cuba.*[3]

JAMES BAMSFORD: *Have the CIA secretly create terrorism in the United States. People would be shot on American streets, bombs would be blown up.*

These are referring to idea number 4, which says in full:

We could develop a Communist Cuban terror campaign in the Miami area, in other Florida cities and even in Washington. The terror campaign could be pointed at refugees seeking haven in the United States. We could sink a boatload of Cubans enroute to Florida (real or simulated). We could foster attempts on lives of Cuban refugees in the United States even to the extent of wounding in instances to be widely publicized. Exploding a few plastic bombs in carefully chosen spots, the arrest of Cuban agents and the release of prepared documents substantiating Cuban involvement also would be helpful in projecting the idea of an irresponsible government.

This is by far the most serious of the nine suggestions in that it actually goes as far as "wounding" Cuban refugees. More ominously, though with no real detail, they suggest a "real" sinking of "a boatload of Cubans"—although they don't say if they would rescue them or not.

Is Ventura's description accurate? At face value it initially seems so. However, the plan is not to "attack certain cities," it's a plan to appear to target Cuban refugees. Ventura is trying to link Northwoods to 9/11, and yet the plan is a far cry from murdering thousands of people and destroying billions of dollars' worth of buildings. The possible wounding of a few Cuban refugees is very different to the carnage of 9/11.

Alex Jones had a more extreme interpretation.

ALEX JONES: The federal government proposed blowing up airliners full of Americans, saying the casualty list in US newspapers would cause a helpful wave of indignation.

ALEX JONES: In the plan they elaborated on how they could 'bomb Washington DC' and blame it on Cuba, attack Marines at Guantanamo Bay using US Army soldiers dressed up as Cubans, or, just like the sinking of the Maine to get into the Spanish–American War, they could blow up a ship, again![4]

Jones is taking the proposals to stage attacks and presenting them as if they actually involve killing people. This is false. There was no plan to actually kill Americans—the proposals to blow up ships and airplanes only involved drone aircraft. Suggestion number 8, one of the more complex suggestions, was:

It is possible to create an incident which will demonstrate convincingly that a Cuban aircraft has attacked and shot down a chartered civil airliner enroute from the United States to Jamaica, Guatemala, Panama or Venezuela. The destination would be chosen only to cause the flight plan route to cross Cuba. The passengers could be a group of college students off on a holiday or any grouping of persons with a common interest to support chartering a non-scheduled flight.

a. An aircraft at Eglin AFB would be painted and numbered as an exact duplicate for a civil registered aircraft belonging to a CIA proprietary organization in the Miami area. At a designated time the duplicate would be substituted for the actual civil aircraft and would be loaded with the selected passengers, all boarded under carefully prepared aliases. The actual registered aircraft would be converted to a drone.

b. Take off times of the drone aircraft and the actual aircraft will be scheduled to allow a rendezvous south of Florida. From the rendezvous point the passenger-carrying aircraft will descend to minimum altitude and go directly into an auxiliary field at Eglin AFB where arrangements will have been made to evacuate the passengers and return the aircraft to its original status. The drone aircraft meanwhile will continue to fly the filed flight plan. When over Cuba the drone will be transmitting on the international distress frequency a "MAY DAY" message stating he is under attack by Cuban MiG aircraft. The transmission will be interrupted by destruction of the aircraft which will be triggered by radio signal. This will allow ICAO radio stations in the Western Hemisphere to tell the US what has happened to the aircraft instead of the US trying to "sell" the incident.

Since Operation Northwoods comes up quite often, it's quite important for you to understand its history, the historical context, the people involved, and the actual contents of the document. The two important things to remember about Northwoods are:

1) It was a series of preliminary proposals that never got even got to the initial planning stages.

2) In none of the incidents were American *citizens* intended to be hurt or killed. Although, there was the potential for Cuban nationals to be harmed, and of course any invasion of Cuba would likely cost hundreds or thousands of lives.

There's a note at the start of the list that makes clear that this was just series of preliminary proposals:

Note: The courses of action which follow are a preliminary submission suitable only for planning purposes. They are arranged neither chronologically nor in ascending order. Together with similar inputs from other agencies, they are intended to provide a point of departure for the development of a single, integrated, time-phased plan. Such a plan would permit the evaluation of individual projects within the context of cumulative, correlated actions designed to lead inexorably to the objective of adequate justification for US military intervention in Cuba.[5]

Basically, it was a collection of ideas intended to be used as a starting point to develop actual plans. None of the ideas appear to have proceeded past this initial speculation. The Joint Chiefs of Staff did not approve of these ideas being implemented. They simply approved of this document as a starting point for discussions.

And we know the ideas were never implemented *because the events described never happened.* If any of the suggestions (beyond the more banal ones, like "spreading rumors") were implemented then they would have made the news. But there's no reports of passenger planes being buzzed by Cuban MiGs, no party of teenagers went missing, no Navy jet was "lost" over Cuba.

Context is of vital importance in understanding the merits of any claim. Let's look at the timeline of events around the time of the Operation Northwoods document.

- 17 April 1961—Bay of Pigs Invasion—a failed military invasion of Cuba undertaken by CIA-sponsored former Cubans. The plan was developed by Eisenhower but approved by JFK.
- November 1961—Decision to implement the "Cuban Project," a.k.a. "Operation Mongoose," a plan of sabotage and propaganda.
- 5 March 1962—The head of the Cuban Project writes a memo headed "Operation Mongoose" requesting a list of pretexts that would provide justification for a US invasion of Cuba.
- 7 March 1962—The Joint Chiefs indicate a desire to develop a Cuban provocation.
- 9 March 1962—Date of the Operation Northwoods memo with the list of suggestions.
- 13 March 1962—The Joint Chiefs of Staff recommend the list of nine suggestions comprising Operation Northwoods be "forwarded as a preliminary submission suitable for planning purposes," assuming "that there will be similar submissions from other agencies."
- May 1962—Plans were suggested to fly U2s unnecessarily over Cuba to provoke an attack by the Cubans.
- 14 October 1962—High altitude photos prove Soviet-made missiles were in Cuba, leading to the start of the Cuban Missile Crisis.

Notice that The Cuban Project and the Operation Northwoods documents are bookended by the Bay of Pigs and the Cuban Missile Crisis. JFK had been anxious all along to avoid a confrontation with Cuba that might escalate to war with Russia. The Bay of Pigs had been framed as a counterrevolution specifically to allow America plausible deniability. After this failed there was a focus on finding adequate pretexts to allow the US to invade Cuba without giving the Soviets reason to respond in kind. Operation Northwoods was part of this, but ultimately went nowhere.

Why didn't Northwoods proceed? The documents provide few specifics, but there are clear concerns on a number of fronts. We might summarize the probable reasons as:

- There was significant risk of being caught. The US wanted to get rid of the communist regime in Cuba, but they did not want war with Russia. They would not proceed unless it was certain of success. Getting caught would have huge repercussions both internationally and at home.
- Many in the government and military simply did not want to invade Cuba, regardless of the validity of the reasons for doing so. Hence there was no point creating false pretexts for something they did not actually want to do.
- Individuals might have been concerned about the morality of possibly injuring Cuban refugees. This concern might have also manifested in a secondary manner where the concern would simply be that the plan would make them look too evil, were it actually revealed.
- Events overtook the idea with the Cuban Missile Crisis in October. An invasion of Cuba now needed no real pretext with the presence of nuclear missiles threatening Florida, and yet there was even less of a desire to do anything that might prompt a nuclear event.

There are important lessons to learn from reading the official documents from around that time.[6] Firstly, that the power structure of the government and the military was (and is) not of one mind. It is made up of people (all men back then) with their own differing experiences, skills, ideologies, and motivations. Many different plans for both sabotage and propaganda were suggested and discussed. The fact that false pretexts (False Flags) were suggested at all indicates that some people were willing to consider them. Many in the military and the government were probably willing to implement them, if they could get away with it.

That bears repeating: *if they could get away with it.* Would you rob a bank of $10 million if all you had to do was click a button, and you were 100 percent guaranteed to get away with it, with no consequences, just $10 million in your bank account no questions asked?

Many people would say yes. But even then, they hesitate—and for similar reasons the Joint Chiefs might have hesitated regarding Northwoods. Is it really 100 percent without risk? Even if there's a small chance of getting caught you are looking at losing everything, and spending years in jail. Unless the assurance comes from God himself (and even then . . .) can you really be sure it's going to work? The risk is unknown. The consequences are severe.

There's other concerns that different people might have. The morality of the situation would put many people off. The consequences of success might even put off a few—we've all heard tales of lottery winners whose lives were ruined after getting the mega-millions. "Don't rock the boat" works for many people.

But many people would take the $10 million. Why didn't Kennedy implement Northwoods? Probably not so much because he thought it was immoral (most of the plans involved no loss of life) but rather from a desire to avoid war with Russia. He knew it was impossible for it to be risk-free, but mostly he did not want to invade Cuba and provoke a war.

The risk did not match the reward. That's the key lesson that you can take from Operation Northwoods, and it's the key way we can take what your friend considers to be an ace card with the claim of a False Flag and turn it around. Sure, you can tell him (quite honestly) people would do a lot if they could get away with it and if there's a big payoff. But Northwoods *was not implemented* precisely because they could not guarantee getting away with it, the risks were bigger than the rewards, and even a successful payoff carried risks of its own.

Ask your friend about their favorite "false flag" theory: what are the risks? What are the rewards? Could the goal have been achieved an easier way? With many proposed 9/11 theories the risks were huge. If one president was incapable of covering up Watergate (a relatively simple burglary and wire-tapping job), then how could another cover up the secret wiring of three huge buildings with millions of explosives, then organize the hijackings of four planes? How exactly do you create a risk-free fake school shooting involving an entire community of fake actors? And for what? A minor shift in public opinion on gun control? A shift that did not actually happen, and actually resulted in an increase in gun sales? What president, what person in any position of power, would risk everything just to get this inconsequential blip in public opinion?

If they bring up Operation Northwoods, seize the opportunity. Go over it with them, see what the plans were, what the risks were, and why they did not do them. Then compare that directly with the theories of your friend.

While Northwoods was never implemented, the use of False Flags or false pretexts for action has certainly been suggested and implemented at the highest levels of government, in the US and around the world. It is quite certain that false pretexts have been used, possibly many times, in the history of US military intervention. As mentioned in the introduction, conspiracies are very real. An incident that many people think as a prime example of this is the event that precipitated the Vietnam War: the Gulf of Tonkin Incident. But like Northwoods, it's an incident that is greatly misunderstood, a card in the deck that can again be flipped around to shed some genuine perspective on what types of false flag are actually plausible.

The Gulf of Tonkin

The two Gulf of Tonkin Incidents seemed like relatively simple things. The USS *Maddox*, a Navy destroyer, was reportedly attacked by North Vietnamese boats on August 2, 1964. A second and larger attack on the *Maddox* and another destroyer (the *Turner Joy*) was reported two days later on August 4. These reported acts of aggression led to retaliatory attacks by the US on the North Vietnamese on August 5 and the passing of the Gulf of Tonkin Resolution on August 7, which then led to full-scale military involvement in the Vietnamese conflict, and hence to the Vietnam War and the loss of thousands of US and Vietnamese lives.

But it's more complicated than that. People thought, on the day, that there was a real attack on August 4. The destroyers took evasive action and fired on what they thought were enemy torpedo boats. But it quickly became clear that there were serious doubts about exactly what was happening. Later it became even more clear that there was no real attack at all. The account given to the American public and to Congress was not an accurate representation of what had happened on either day, but especially not on the second day.

Even President Johnson was not fully informed as to exactly what was going on. On the evening of August 4, he addressed the nation and said:

> As President and Commander in Chief, it is my duty to the American people to report that renewed hostile actions against United States ships on the high seas in the Gulf of Tonkin have today required me to order the military forces of the United States to take action in reply.
>
> The initial attack on the destroyer Maddox, on August 2, was repeated today by a number of hostile vessels attacking two US destroyers with torpedoes. The destroyers and supporting aircraft acted at once on the orders I gave after the initial act of aggression. We believe at least two of the attacking boats were sunk. There were no US losses.[7]

But despite Johnson's resolute tone and his description of events with specifics like "a number of hostile vessels" and "with torpedoes," he was far from certain about what had happened. This was partly because of the "fog of war," the difficulty in getting accurate information from the battlefront. But it was mostly because the initial reports of an attack were wrong, but Johnson did not find this out until later.

To understand these events, we need to understand the role of Robert McNamara, Johnson's Secretary of Defense and the man most influential in the escalation of the Vietnam situation. McNamara was a trusted advisor to Johnson, who relied on him for intelligence briefings regarding what was going on in Vietnam. They spoke frequently on the phone, and many of those conversations were recorded and are now publicly available. What this particular conspiracy boils down to can be summed up as "what McNamara did or did not tell Johnson."

The first reports that came from the Navy ships in the Gulf of Tonkin were unequivocal that the attacks were taking place, and that they were being attacked by North Vietnamese torpedo boats. McNamara was being kept continually up to date on the situation, and since Captain Herrick of the Maddox initially thought he was under attack, so did McNamara.

McNamara was of the opinion that the best course of action in Vietnam was to bomb the North Vietnamese to cripple their military and industrial capacity and force a resolution to the conflict. Johnson was unwilling to take this step unilaterally and insisted that they wait until the North Vietnamese had attacked them. The very limited attack by what turned out to be a small boat on August 2 was insufficient provocation. But now there were two US destroyers being attacked by torpedo boats. Clearly this was the larger scale provocation that McNamara had hoped would happen.

But five hours after the "attacks," Captain Herrick had reviewed the incident and began to have serious doubts as to what actually happened. He sent this message to Washington, where it was received at 1:27 p.m., Washington time:

Review of action makes many reported contacts and torpedoes fired appear doubtful. Freak weather effects on radar and overeager [sonar operators] may have accounted for many reports. No actual visual sightings by Maddox. *Suggest complete evaluation before any further action taken.*

Basically an "oops, we goofed, there was no attack, don't do anything" message. But what happened next was McNamara ignored that message, almost pretending like he never got it. He did not immediately inform Johnson, and instead went ahead organizing and then executing an order for retaliatory air strikes against North Vietnam, signed off by President Johnson, thus making the conflict real, and laying the foundation for the Gulf of Tonkin Resolution, and hence the full Vietnam War.

Public suspicion about this story began almost immediately, with many contemporary articles casting doubt on if the August 4 incident had ever happened. Importantly though, there was never any real doubt about the first incident on August 2. A common simple mistake made by conspiracists has been to describe the "Gulf of Tonkin Incident" as one single event, when in fact there were two separate events. Many conspiracists are very unfamiliar with the details and may be surprised to learn of the first attack.

In 1969 Joseph Goulden's book *Truth is the First Casualty* brought the issue to the attention of a broader audience. Goulden presented convincing evidence from eyewitness accounts that the second attack never took place, and that details of the first attack were somewhat overblown.

But the most detailed revelations regarding what really happened eventually came from the National Security Agency (NSA)—the government organization tasked with intercepting and interpreting foreign communications. The NSA had long stuck to the official story at the time, claiming there was good SIGINT (Signals Intelligence—generally radio traffic and coded communications) evidence to support both the planning and the actual events of second attack. But with a 1998 internal document that was released under the Freedom of Information Act in 2006, the official position changed.[8]

Written by Robert J. Hanyok, a historian and intelligence analyst at the NSA, the document was titled "Skunks, Bogies, Silent Hounds, and the Flying Fish: The Gulf of Tonkin Mystery, 2–4 August 1964."[9] It goes into great detail, not only about what was observed on the boats, but also the SIGINT of intercepted North Vietnamese communications from those days and the days after.

The conclusion was what everyone basically knew by then, that the second attack, the one referenced by Johnson in his call for action that led to the Vietnam War, did not happen. The conclusion that the attacks were real (held by the NSA long after the actual events) was formed by "a compound of analytic errors and an unwillingness to consider contrary evidence." Not only did it not happen, but Hanyok concluded that members of the intelligence community deliberately withheld information that suggested the second attack was illusory.

What we have here is a conspiracy very low down on the IHOP scale—perhaps a small step above a simple "glad it happened" explanation. Nothing was fabricated, there was an actual event of mistaken radar returns and imaginary engagements that Hanyok's analysis shows were believed to be real by the people there at the time. So real in fact that many continued to insist that they saw something for years afterwards. What seems to have happened is that McNamara took advantage of the confusion surrounding the event and deliberately misled Johnson regarding the certainty of the reports.

What do you tell your conspiracy friend who raises the Gulf of Tonkin as an example of a false flag? You should embrace the opportunity to talk about real corruption and duplicity in government. Yes, the American involvement in Vietnam was greatly escalated on what was essentially a false pretext. But it was not a created incident. McNamara knew a second incident was needed after the August 2 incident, and perhaps he was deliberately sending ships into harm's way, certainly he seemed to be hoping there would be a second attack. But the August 4 incident was a genuine mistake, not a falsified incident. McNamara leapt upon it when he thought it was real, and then ignored suggestions and increasing evidence that it was not. He distorted, delayed, and misled. He exploited a mistake, but he did not create a fake incident.

This type of distortion and misleading by power factions within the administration is nothing new, nor is it anything particularly surprising. People want particular outcomes, and if a certain interpretation of events can help lead towards that outcome then they will try to spin that interpretation with the vigor of a devil's advocate.

The most obvious examples of this type of spun-justification come from the events that follow the 9/11 attacks. Here there clearly was a real incident, but Bush and Cheney wanted to invade Iraq, not Afghanistan, and certainly not Saudi Arabia (where most of the hijackers were from, and quite possibly where some of al Qaeda's financing originated). While 9/11 gave them a mandate to strike back, they could not immediately invade Iraq in response. Significant spin and distortion were required to pull that off, most notably the magnification of any evidence regarding Weapons of Mass Destruction, and the ignoring of any evidence that suggested those weapons did not exist. That cherry-picking is similar to what McNamara did with the evidence of the August 4, 1964 attack, but on a slower and a grander scale.

Conspiracy theories grow largely in the absence of a clear picture of what actually happened. We don't often get a clear view of high-level discussion regarding national security for obvious reasons. But that secrecy diminishes over time. The McNamara Tonkin incident is old enough, and declassified enough, to give us a unique look into "how the sausage is made." While normally nobody else is in the room where it happened, here we have the actual tapes of McNamara first not telling Johnson, and then telling him, and then

Johnson getting annoyed for not being told. The more a conspiracy theorist understands of the actual machinations of power, the less likely it is that their false interpretations will stick. You should encourage your friend to "do their own research" on the matter. Have them find out what happened on August 2 and August 4, what Johnson knew, what McNamara knew, and how and when they knew it.

This is even one case when doing YouTube research can be quite fruitful. A search for "McNamara Tonkin" will give you several interesting results, including several extended interviews with McNamara discussing the incident and the broader Vietnam situation. Beware of videos that only give you short clips of the phone conversations, the full versions with accurate transcripts are available at the National Archives of GWU.[10] Listen to them with your friend if possible; read the transcripts in full.

Note McNamara's eagerness to bomb the gasoline depots, note how he assumes that there's going to be a second attack at some point, and how he's already making plans for it. Then get the backstory on that, read Hanyok's article, and understand *why* McNamara anticipated a second attack—it was the SIGINT that mistakenly predicted one. The conspiracy videos often highlight the fact that McNamara said "second attack" before any second attack took place. At face value this seems to be evidence that McNamara had planned a second attack. Then (the conspiracy logic goes) since there was no real second attack, McNamara must have planned a fake second attack—the "false flag." But what he was referring to was the fact that intelligence reports had concluded there would be an attack, as explained in great detail in Hanyok's analysis for the NSA.

When your friend brings up Tonkin he will likely assume you have never heard of it. He sees things like Tonkin and Northwoods as special knowledge that only he possesses. Your first challenge is actually going to establish the common ground. You have to find out how much they know, and then you've got to show him how much you know. Physical objects are good if you are discussing things in person. Show them a printout of the Northwoods documents, or the Tonkin Johnson/McNamara transcripts—ideally with key portions highlighted. Even if you are not meeting in person it's still sometimes useful to show them a photograph of yourself holding an actual document. Links to the document are useful, but in the midst of a conversation often are just throwaways that go nowhere. You can have the best links in the world, but it takes effort to click and read. Show them the actual document and ask them, "I have read this, have you?"

The best common ground is one shared against some common foe. We see in the historical cases of Northwoods and Tonkin a willingness of people in power to exploit events in order to use them as pretexts for action, sometimes even as pretexts for war. We see in Northwoods a willingness to consider false and fabricated pretexts—very real False Flags. We know there is corruption in the halls of power. We know that people who wield power have their own individual self-interests and ideologies as their primary driving motivations, and that often (but not always) those self-interests and ideologies do not mesh well with the common person's.

We also see in these historical examples the limitations of fabricated events. The Joint Chiefs of Staff are quite clear in Northwoods that they would prefer a real reason to invade Cuba. Their first thought was to provoke an action by Cuba, rather than try to pretend one had happened.

> *Since it would seem desirable to use legitimate provocation as the basis for US military intervention in Cuba, [a plan] could be executed as an initial effort to provoke Cuban reactions. Harassment plus deceptive actions to convince the Cubans of imminent invasion would be emphasized. Our military posture throughout execution of the plan will allow a rapid change from exercise to intervention if Cuban response justifies.*[11]

Prod the nest until the hornets get angry enough to sting, then justifiably burn it down. Arguably this is what McNamara was doing in Vietnam; he probably wanted the North Vietnamese to attack at some point, but he wanted a real attack, a "genuine pretext." Perhaps he arranged the patrols to maximize the chance of such an attack, but what actually happened was an accident, a mistake that he exploited and then was forced to go along with.

While truth is often said to be stranger than fiction, in the conspiracy world the fiction is generally a lot stranger than the truth. But truth is still complex and messy. We can never say exactly what happened in the Gulf of

Tonkin, we will never know what was in the mind of President Kennedy when he rejected Northwoods. But there's a lot we can know about both situations, especially Tonkin. There's a lot that most conspiracy theorists are simply unaware of, and it's by shining a light on these rather mundane truths of the fog of war, of spin, of event exploitation, that we get the clearest picture of what actually happened. Northwoods was spitballing ideas that did not go anywhere. Tonkin was a mistaken battle against radar ghosts, exploited to push an agenda. Once your friend sees these things, sees the actual facts in detail, then there's less need in his mind for conspiracies.

Do not be disheartened if after all your attempts to show the reality behind historical False Flags your friend still turns around and says, "See what the government does." These things take time. Focus not so much on trying to change their mind as on filling their minds with as much factual information as possible. Emerging from the rabbit hole rarely happens as a sudden realization. It's a gradual awareness of reality that eventually crowds out the fantasy. People who think that the shootings at Sandy Hook and Las Vegas were staged are very deep down the hole. They have huge gaps and distortions in their understanding of the world. Debunking solely one point is not going to get them out. A substantial amount of illumination is required, and often a substantial amount of time to let them look around and contemplate their newfound knowledge. If you are simply getting them to honestly listen to you and to look at shared factual information, then you and your friend are taking steps in the right direction.

False Flag Bombings and Shootings

On April 15, 2013, I was in a hardware store when I received a text from a friend who never texts me. It said:

> *I realize tens of thousands of people run the Boston marathon, but FYI if you haven't heard there is some incident at the finish line that happened very shortly ago.*

Then my wife texted:
COME HOME NOW!

My wife's father was running the Boston Marathon, as he did every year. Two bombs had just gone off near the finish line and there were casualities. We had no way of getting in touch with him and we spent the next few hours glued to the TV trying to spot him in various shots. We knew he was about due to finish, and at first we were worried that he might even be the now-iconic runner who stumbled and fell as the explosions went off just behind him.

Disasters, mass shootings, and bombings are personal to different degrees to everyone. It was very real to the people who were there that day, to the people who ran the race and witnessed what happened. My father-in-law was just five minutes away from the finish line when the race was shut down. The police corralled the runners in a safe area for a while, and eventually he found a phone and let us know that he was safe. It was real to him, and it was real to us.

Most people who watched the footage of the bombing had nobody there, but they had no problem accepting it as real. They could see there were hundreds of people lining the road, tens of thousands of people had already run over the finish line. The explosions, the shock, the responses, the blood, all were very real.

Yet a small group of people insisted it was not. Scouring over the footage they looked for any little inconsistency, anything they could not immediately explain, and held it up as "evidence" that the entire event had been faked.

These are the same people who also thought that the Sandy Hook school massacre was faked and would later think that the Las Vegas shooting was faked. We know they are the same people because this type of "false flag" conspiracy very rapidly becomes a mental trap. Once they accept that one extraordinary event is fake, then that means the anything could be faked and (in their mind) probably is. It's a dark corner deep in the rabbit hole from which it is very difficult to escape.

What makes it particularly difficult is that the adherents to these theories become adept in incorporating new information into their narratives. While there may be some who can be swayed by the usual approach of pointing out their errors (the spotlight approach), the problem is often more fundamental—an entire worldview that is based on everything being an illusion.

What is needed is more of a floodlight. They have to gain some real perspective on how the world really works. Discussing the mundane realities of Northwoods and Tonkin might help. But you're also going to have to try to bring them a broader-based perspective on other matters. The litany of ridiculous claims for such things is endless, but there's a few key things that come up again and again in various forms, and which can (sometimes) be addressed with appropriate perspective.

Changes in the Initial Story

During chaotic events, what is known by the media and law enforcement is rapidly changing, and often based on second- or third-hand accounts. With time the story gets more accurate, but this means that it changes as clarity is achieved. To some people this seems suspicious.

The most common version of this is there being reports of multiple gunmen in the single-shooter situations. With Sandy Hook there was a "man in the woods" (at least two, later identified as a parent and an off-duty policeman).[12] With Las Vegas there were reports of shots fired from a different location (a distant flashing light reflecting off a window),[13] with Pulse there were anonymous internet posts claiming a second shooter was present, later proven false.[14] There were also a variety of claims of second (and third, maybe even fourth) shooters at the JFK murder.

The main reason these stories take hold is that the police initially don't know how many shooters there are. They always go into a situation *by default* looking for additional shooters even if the primary suspect is dead or captured. The media reports that they are looking, and the false story gains some legs.

You can point this out, and you can also point out that *none* of these reports of second shooters actually panned out. But perhaps the most valuable perspective is to be gained in looking at just how (un)reliable eyewitnesses are. Research, and practical experience, has shown time and again that witnesses to chaotic and violent events usually have very different recollections of the details.[15]

False Injuries and Blood

The aftermath of the Boston Marathon bombing revealed a variety of injuries. People had their legs blown off, bones were broken, shrapnel was embedded in limbs, and blood was everywhere. People raised all kinds of specious objection to what they saw. The blood was "too red," people walked when they should not be able to, injuries seemed inconsistent with later photos, limbs appeared to be at the wrong angle.

Similar objections are raised with respect to other events. When a TV reporter was shot and subsequently ran away from the shooter, people asked how she could still run after being shot. When JFK was shot from the back, people were confused because his head jerked back and to the left.

These types of claims are simply based on misunderstandings of how the human body works, and limits in the personal experience of those believing the claims. Blood from a little cut in your finger is a dull red, but freshly spilled blood from an artery *is* actually bright red due to the high oxygen content. In many action films when people get shot they just drop dead, but in reality gunshots are frequently non-fatal. People can get around with the most horrific injuries. Later surgeries (for example, to repair an injured knee) often temporarily incapacitate people more severely than the event that caused the injury.

Inappropriate Emotions

When the father of one of the young Sandy Hook victims briefly laughed and smiled at his daughter's funeral, people said he forgot he was supposed to be acting sad. When a man who took in some of the kids fleeing the school burst into tears when relating the story, people said he was overacting.

Belief in these claims seems to be based on a lack of life experience, and in particular the various ways in which people deal with loss or traumatic events. People do actually laugh and smile at funerals. Not all the time obviously, but it certainly happens. A recent funeral I attended after the sudden death of a dear friend was one of the saddest days of my life, and yet was also filled with the joy of friendship and happy memories. There were tears, and gut-wrenching grief, but there were also jokes, smiles, and laughter. Humans are complex—and there aren't any firm rules for how an individual is supposed to emotionally respond to trauma. This is how people are.

How do you show this? You can relate your own experience, you can show videos of funerals, you can show them written accounts, and show psychologists' research into dealing with grief.[16] Hopefully you can reach them, but for some it might simply take the inevitable gaining of life experience that they will get as they age.

People Looking Like Other People

A common claim is that "crisis actors" are used to stage these events, and that the same actors keep showing up. This claim is invariably backed with a photo comparison that show two people that look similar. It is invariably debunked by showing higher resolution images of those two people from different angles that demonstrate conclusively that they are different people.

There are two aspects of perspective that you can bring here. Firstly, it's not that hard to find people that look alike if you try hard enough. Even within a relatively small group, like celebrities, there's many that look nearly indistinguishable in some photos. Katy Perry looks like Zooey Deschanel, Bob Saget looks like Stephen Colbert. There's lots of collections of celebrity doppelgängers online; show them to your friend.

Secondly, these claims are invariably shown to be wrong. If your friend thinks some of the people are the same, then they likely believe some classic examples of such claims that have actually been soundly debunked with high resolution photos. Find out which ones they are, look them up, and show them.

Web Pages Created before the Event

Sometimes you'll see some news story, fundraiser, or tweet that seems to be timestamped before the actual event to which it refers. There's often very rational explanations for these anomalies, but they can be hard to explain as they are occasionally a little technical. The best approach is to show examples of them happening in events that are not disputed.

A common mistake is to report the time stamp of a tweet in a different time zone to when it happened. After the Boston Marathon bombing (which happened at 2:50 p.m. local time), a certain tweet looked different on the East Coast and the West Coast.

FIGURE 13: A tweet from after the Boston Bombing viewed from two different time zones. From the West Coast it looks like the tweet preceded the bombing, when it was actually after it.

The West Coast image with the 12:53 time stamp was spread around as evidence that the bombing was a controlled explosion gone wrong. The simplest thing to do here is show them how to change the time zone on Twitter and show them this tweet (or whichever one they were suspicious about) in the original time zone.

There's other ways times and dates can be misleading. When you do a Google search for an event (like "Sandy Hook Shooting") you often get results that Google says are from *before* that event. This is just a glitch in Google and you can demonstrate this to your friend by searching for stories about something neutral but unique. For example you could look up the 2016 film *Zootopia*, about which there's dozens of web pages that are dated prior to 2010. In Google, just click "Tools," then "Any Time," then "Custom Range," and enter 2011 in the "To" box. The supposedly predated pages will appear. Disney and Google aren't coordinating to cover-up the true release date of *Zootopia*, it was just a technological glitch.

This is just a small sprinkling of the wide variety of specious claims that pepper false flag mass casualty conspiracy theories. A comprehensive debunking of even one such conspiracy theory could easily fill its own book. But what they all have in common is that their claims do not hold up to investigation. Show your friend these common mistakes, and later ask them to add "debunked" or "Snopes" to their internet searches, just to get the other side. Show them enough things that are wrong, and eventually the balance will shift from automatic acceptance of alternative narratives to more cautious and reasonable fact-checking. It will probably take some time.

False Flag Summary and Resources

Key Points to Convey to Your Friend

- They should not blindly trust the government or any grossly powerful organization, and you don't trust the government either.
- "False Flag" refers to a real type of event. But it does not follow that all suggested false flag theories are correct.
- False flag theories are suggested after every mass casualty event. It's more of an automatic response than an evidence-based theory.
- Operation Northwoods was not an operation; it was a list of suggestions that were never implemented. It's also pretty much the only example of such a thing.
- The Gulf of Tonkin highlights an actual misleadingly spun incident. But it was an exploited mistake, not an invented attack.
- All the claims of evidence for false flag mass casualty events, so far, simply do not stand up to detailed scrutiny. There's a lot of them, and they are all, invariably, wrong. Keep this in mind the next time one comes along.

Additional Resources

- Metabunk's Sandy Hook and Boston Bombing forums—**metabunk.org /sandy-hook.f24 and metabunk.org/boston.f27**
- Snopes Analysis of "Sandy Hook Exposed"—**snopes.com/fact-check /sandy-hook-exposed**
- Snopes analysis of Las Vegas shooting—**snopes.com/fact-check /las-vegas-shooting-rumors-hoaxes-and-conspiracy-theories/**
- Operation Northwoods document—**nsarchive2.gwu.edu/news/20010430/**
- Gulf of Tonkin audio archive—**nsarchive2.gwu.edu//NSAEBB /NSAEBB132/tapes.htm**

CHAPTER TWELVE

Richard – Drawing the Line at Sandy Hook

Richard is a young man from Chicago. He got into conspiracy theories in college, at a troubled point in his life.

When I look back in hindsight at everything that happened that made me get in so deep, I think this might just be a symptom of being a young person. Like that stuff you see on college campuses, all that stuff going on, searching to be a part of something. I was also going through a pretty angry time in my life. In high school my father passed away, my mother struggled a lot after that, it was kind of a tough time. My grades suffered, I still made it into college, but things just really weren't going very well.

I spend a lot of time online, going down rabbit holes at two in the morning. I just started stumbling upon certain videos, certain documentaries. It was like a real-life thriller suspense kind of thing, you know, getting that dopamine drip. It's almost like you are watching a movie unfold with all these twists and turns and mysteries. But it's real life, and I'm sitting there on a laptop going down this rabbit hole and I'm clicking on more videos, getting more drawn in.

The 9/11 attacks happened when Richard was just ten years old. This was a very traumatic event for him and the memories of seeing it happen on TV are still with him. He was naturally drawn by fascination into the 9/11 Truth conspiracy rabbit hole. It was not long before he encountered the works of Alex Jones, a whole new series of rabbit holes opened up, and he started to feel like part of an important community.

You start feeling like "we're the awake ones." It draws you in with the facade of "you're being lied to, you're being brainwashed," and the Alex Joneses of the world want to give you the real business. Now you're in there, you're part of the "awake" group, you're special now. What's funny is they do have some nuggets of truth about weird things that happened in the past. They throw a wide net, it's like information overload, ranting for thirty minutes, too many things to follow, and then hitting those small nuggets of truth that kind of drag you along. It traps you without thinking, as there's no way you'd think that you are part of that bubble of dishonesty that you criticize other people for, so you don't really see it happening.

Richard got deep down in the rabbit hole, and yet like most people he settled more or less on one position on the conspiracy spectrum. He believed many 9/11 theories, Chemtrails, and some kind of "New World Order" conspiracy, and of course the JFK assassination was a given. But he drew the line at the more extreme things. He'd always liked space and astronomy as a kid, and perhaps that helped him steer clear of the Moon Landing Hoax and Flat Earth theories.

People often get out of the rabbit hole by questioning more extreme beliefs, and that's partly what happened to Richard. But the thing that started him questioning was rather unusually low down on the spectrum: vaccines. Alex Jones is a big promoter of the anti-vaccine theory, so Richard naturally believed that vaccines were just some Big Government/Pharma plot. Then his sister went to medical school.

She got a couple of years in, and one day we were home at dinner, and she brought it up. I didn't want to get into a whole thing, but I did say something and then she just really eviscerated some of the notions I had about vaccines. That kind of set off the first alarm. I thought, wow, this is a person who has dedicated her entire life for the last couple of years into looking into this stuff, and what she knows just shatters what I thought I knew. And she wasn't even a professional yet, she was just a med student.

His sister's information about vaccines got him thinking, but the topic that really started him on his journey out was just on the other side of his line of demarcation. He was previously willing to consider false flag shooting theories as plausible, but then, with the new thoughtfulness prompted by his sister, he looked more closely at the evidence.

The Sandy Hook thing, the shooting in Connecticut of first graders, that was something that was very startling. That was the first time I saw 100 percent that Alex Jones and that community were being very dishonest. Like they said the FBI never recorded the deaths for Sandy Hook. But then I found, no, it was in this separate file, "FBI Miscellaneous," and it was clearly recorded, and clearly there. So Alex Jones and those people were clearly being dishonest.

Richard frequented some online communities that would discuss supposed False Flags. But when he brought up this "death files" mistake he was shunned. People either ignored him or they insulted him, criticizing him for being a "sheep" repeating government propaganda. Nobody really addressed what he was saying, they just rejected it out of hand.

Experiences like this started to click into place in Richard's mind, starting what he described as a "domino effect," where one thing led to another. He became aware of a "gang-like" mentality in the 9/11 Truther community and saw further examples of people simply rejecting evidence without considering it.

There was just something rubbing me the wrong way about the way these people were going about this stuff. The reason I got into this whole conspiracy thing was because I wanted to be this open-minded guy. So how could I be? I had to at least consider other sides here. So, I started to see where they had been dishonest, like in the 9/11 steel beams argument.[1] I'll never forget this video this blacksmith did,[2] he kind of explained how beams can weaken at something like only 800 degrees and did a demonstration on it.

Something of an anti-rabbit hole opened for Richard and he started to look at more 9/11 debunking videos about the internal structure of the World Trade Center towers, and videos showing just what an inferno it was inside Building 7, and it all started to make sense. The dominos started to fall, and after a while he stepped away from looking at anything on either side, not really wanting to fully admit he'd been duped. It wasn't long before the last penny dropped.

I was in my car, and on the radio the subject of 9/11 Truth was brought up, so immediately I'm glued to the radio. These Truthers kept calling in to the show and the "facts" they were spewing were just the same things I'd heard before, and by this time I knew that most of them were just complete nonsense. It wasn't long after that that I thought: this is all over.

There wasn't a single "aha" moment for Richard; it was, as he says, "a little bit here, a little bit there." There are some key moments like his sister explaining vaccines, the Sandy Hook Truthers ignoring actual evidence, the dishonesty of the repeated 9/11 arguments, and then finally, after spending enough time with "the other side," he accidentally revisited the claims of the 9/11 Truthers with a clearer, better informed mind and saw those claims for what they really were.

Richard recognized a political bias in people like Alex Jones. As he saw it, Jones was very quick to call any school shooting or other shootings by Americans a "false flag," but when there was an act of jihadi terrorism Jones simply accepted it as a genuine event and attacked Islam.

The thing that really pulled at my heart string was the Sandy Hook parents being told that they were liars, actors. There was one parent, my heart just broke for him. He wanted to prove that his daughter's death was legitimate, so he thought he'd just show the death certificate and put a stop to it. So, he posted the death certificate and they were like, "Oh, it's the wrong color ink." They were just finding these little things that they could turn into inconsistencies.

Richard does not have much time for conspiracy theories now that he's gotten out, but he has spent a little time debunking something he never actually believed in, the Flat Earth theory.

I'm one of those foolish people who have tried to go on the Flat Earth Society message boards. Just the complete absurdity of it has drawn me in just out of intrigue, I think. I thought I could just go in there and ask them certain things about the Coriolis effect or eclipses. I've found the Flat Earth people take scientific theories and change them and make them their own. With the eclipse recently, they even tried to spin that as proof of the Flat Earth theory.

I asked Richard about online resources that might have helped along the way.

It's tough because when you are in that mindset you have all this great information online, but you can just choose to shroud yourself in whatever vantage point you want to, and only hear that point. Going online and researching stuff did help the dominos fall, but I think there has to be a will to partake in that research.

I didn't really go with the "everyone's a paid shill" thing or anything like that. But I did think the people on debunking sites were either brainwashed, or just wrong in misrepresenting facts. Their minds were fixed in a certain way.

I remember [when encountering debunkers] hearing information then feeling in my gut like this might be the beginnings of being challenged, and I would run to my "safe space," like googling information from Truthers that would help me feel better. I didn't entertain that side of the argument, and I chose to shroud myself in information that lined up with my views.

Sometimes I was reading something, or seeing something, and out of frustration I was getting angry at what I'm reading and what I'm seeing. Then I'd run to a YouTube video, or I'd just aimlessly type in whatever theory is being brought forth and add "debunked" and find something to keep on the tracks of the Truther movement.

I try now to always look for the most rational explanation. The one thing that I've really gathered from my experience is that all these conspiracies require a big leap of faith, but usually what I find is the simplest explanation is usually true, and you can find A + B = C without a leap of faith.

Richard was helped out of the rabbit hole by his friends (his sister in particular), but it took years to happen. If your friend seems stuck down the rabbit hole, then don't give up. Maybe it will take years, maybe you won't see any progress for a long time. But people find their way out with help, so keep helping.

CHAPTER THIRTEEN

Flat Earth

The Flat Earth conspiracy theory is fundamentally a simple idea: the Earth is flat, and this fact has been covered up by a powerful elite who are pretending that it is a globe. There are some variants, but the most common theory is that the Earth is a flat disk with the North Pole in the center. Instead of Antarctica being a large continent at the South Pole, it is proposed to be an "ice wall" that encircles the edge of the disk.

This theory is firmly at the extreme end of the conspiracy spectrum. It's an extreme conspiracy theory because the belief requires that you accept that the entire space program is a fake, designed to cover up hundreds of years of an even deeper scientific conspiracy to hoodwink people into thinking the Earth is round. You must also accept that GPS does not work by satellites but from radio towers, flights between Australia and South America are fake, the Sun sets via a bizarre interpretation of "perspective," astronauts are all liars, and every image of the Earth from space is doctored.

To the person first looking into this theory it can be difficult to believe that the proponents actually take themselves seriously. Many of them do not. Many Flat Earth popularizers are simply doing it for fun, or to make a philosophical point about people's over-reliance on the authority of science. But there are also people who *do* take it very seriously. Many do so for religious reasons, feeling that a literal reading of the Bible (or sometimes the Koran, or other religious texts) indicates the Earth is flat. Increasingly though there are people, usually young or otherwise easily convinced, who believe that the Earth is flat because they have seen what they think is compelling evidence in YouTube videos.

The religious Flat Earth beliefs are largely faith-based, and so not suscepti-ble to reason. If your friend believes that the Earth is flat *only* because he thinks the Bible tells him so, then unfortunately you are unlikely to make much prog-ress. But even the more religious Flat Earthers attempt to provide scientific proofs of their theory. Some very religious people even claim not to be reli-giously motivated in this topic, and to be approaching it from a purely scientific viewpoint. You can usually make some progress by looking at the science.

While the number of true believers in this theory is quite small, it's still well worth looking into. People who are interested in less extreme theories (like 9/11 controlled demolition) will be upset that I included it in the book. My intent is not to equate the two in terms of being equally extreme, because they generally are not. But the way you'd approach debunking Flat Earth is similar to the way you'd approach debunking other false theories. I would hope that if there are 9/11 Truthers or Chemtrail believers reading this to help debunk a friend's Flat Earth theory, then they might at least consider for a moment that similar scrutiny be given to some of the claims that are just on their side of the line.

To understand a conspiracy theory, it's very useful to understand its history. To that end I highly recommend the book *Flat Earth: The History of an Infa-mous Idea*, by Christine Garwood.

The book first details the history of ideas about the shape of the Earth, going back to ancient times. But the focus is on the Flat Earth movement that sprang up in the late 1800s, a surprisingly similar movement to our current YouTube-driven one. An integral part of the story is the role of skeptics and debunkers who attempted to address the issues back then. There were many points in the book at which I literally laughed out loud with recognition. Not only at situations repeated recently in Flat Earth debunking, but more gener-ally at how the events surrounding the Flat Earth debate mirrored the debates over more modern conspiracy theories like "Chemtrails."

The book opens with a prologue: "The Columbus Blunder," describing the origin of the misconception that Columbus proved the Earth was round.

Even in 1492 the rotundity of the Earth had been known for thousands of years, and very few educated people doubted it. The modern misconception came about from a colorful account of Columbus' life written in 1828 by Washington Irving (author of *The Legend of Sleepy Hollow*).

The story of Columbus creates the backdrop for the first chapter: "Surveying the Earth," where Garwood details the evolution of ideas about the shape of the Earth. The shift in thinking from flat to round is identified as being about 2,500 years ago, in the time of Pythagoras, then Plato and Aristotle. A key thing to explain to your friend is that the shape of the Earth was discovered (and proven) thousands of years ago.

Subsequent chapters are character portraits, mini-biographies of the key figures in the Flat Earth movements. Garwood details the efforts of professed believers—some genuine, some joking, some possibly charlatans. Interwoven are the stories of those opposed to the spread of the false idea, the incredulous and amused public, the debunkers, the people trying to explain things to their friends, back in the 1880s.

The real story begins with Samuel Rowbotham, the socialist manager of a commune in Cambridgeshire, England. Seemingly always an independent and contrary thinker, Rowbotham took advantage of the length and straightness of a local canal to try to determine the curvature of the Earth. By his own account he was unable to detect any, and he quickly became convinced the Earth was actually flat. Simultaneously with this discovery he became convinced that this was also exactly what the Bible described.

Rowbotham discovered he had a talent for persuading people. He began to make a living by selling pamphlets and books with his ideas about the Flat Earth, and by performing lectures which people would pay to attend. He developed the idea of "zetetic" thinking, a concept that is key to understanding the mindset of the Flat Earth believer, both old and new. Pure zeteticism is essentially a form a scientific skepticism where ideas are believed only if you can personally verify them. It is perhaps the ultimate form of "do your own research," where not only do you have to research the evidence for an idea, but you also have to research the very foundation of science itself.

Very little has changed in the Flat Earth arena since the time of Rowbotham. His most popular book, *Zetetic Astronomy,* contains many of the exact same claims of evidence and "proofs" of Flat Earth that are seen today in other Flat Earth books published by supposedly more modern proponents like Eric Dubay. The lectures carried out in church halls are the direct equivalent of YouTube videos about Flat Earth that are still being produced at this very moment.

Even in 1864, an analog version of the internet existed with ink and paper. Communication was much slower. Email was real mail, and discussion forums and comments sections were the letter pages of newspapers. The exploits of "Parallax" (as Rowbotham was known) were often covered by local newspapers, and the letters that followed a Parallax lecture bear a striking similarity to comment threads that might follow a Dubay video. Garwood describes these letters:

> *[In 1864] the paper's correspondence pages were crammed with letters from irate citizens of Plymouth, many of whom were disgusted by zetetic exploits in their town. Among those most appalled were amateur astronomers and local seamen, who wrote in droves complaining about Parallax's "foolish assertions" and his attempts to mislead the public about the most fundamental scientific facts. Keen to make amends, they offered a series of proofs for rotundity, from circumnavigation to the curved shadow of the earth during an eclipse of the moon. One sailor, from a naval and nautical school, even felt it necessary to add that during twenty years of voyaging he had never seen the ice barrier that was supposed to surround the disc-shaped earth, and Parallax's claims to have observed boats at great distances on rivers and seas were impossible unless his eye had been elevated far above water level. Amateur astronomer James Willis agreed, declaring that as Parallax had posted himself as a teacher, he should be willing to replicate his experiments openly for all to see. This drew a response from Parallax who declared, on 6 October, that he was ready, willing and able to, "do battle, inch by inch" with his Newtonian opponents "and upon their own ground."*

Rowbotham continued in this vein for a number of years, others came after him doing the same thing. The reactions were the same—incredulity and derision from the popular press, but inevitably enough people thought he had a point (or they were just entertained enough) that a movement was formed.

Garwood describes a variety of nineteenth and twentieth century followers. There's John Hampton, who infamously refused to accept the result of a wager made with famous scientist Alfred Russel Wallace, and harassed him for years in an 1880s form of cyberstalking conduced largely via letters. Then there's Lady Blount, an energetic lady of leisure who took up the mantel of "Parallax" (d. 1894) by forming the "Universal Zetetic Society" in 1893. The book ends with Charles Kenneth Johnson—the founder of the International Flat Earth Research Society of America. Significantly hampered by the space program, with photos of the Earth from space and live TV shots from the Moon, as well as other problems caused by international air travel, and a general public understanding of time zones, the Flat Earth movement had necessarily moved to an even deeper rejection of reality, now claiming that the entire space program was fake.

> "It's nothing more than a piece of clever stage-managed science-fiction trickery." NASA and world leaders knew that the earth was flat, but they had launched the $24-billion-dollar space hoax as a "scientific plot to hoodwink the public." From Johnson's perspective, there was no alternative explanation: it was impossible to orbit a Flat Earth, rockets could not penetrate the firmament of heaven, and such feats were unnecessary because information about the universe and its creation was laid out in the Book of Genesis.[1]

Garwood's history goes up to 2001, with the death of Johnson presented as the tail end of the decline of the Flat Earth movement, finally fading away under the weight of science and reality. But Johnson died four years before the advent of the greatest vehicle for spreading Flat Earth and other nonsense theories in history: YouTube.

Modern Flat Earth

While YouTube began in 2005, it did not immediately become an amplifier for the Flat Earth belief. If you look at the Google Trends line going back to 2004, interest in the Flat Earth idea was in a slow decline, halving in popularity from 2004 to 2014.

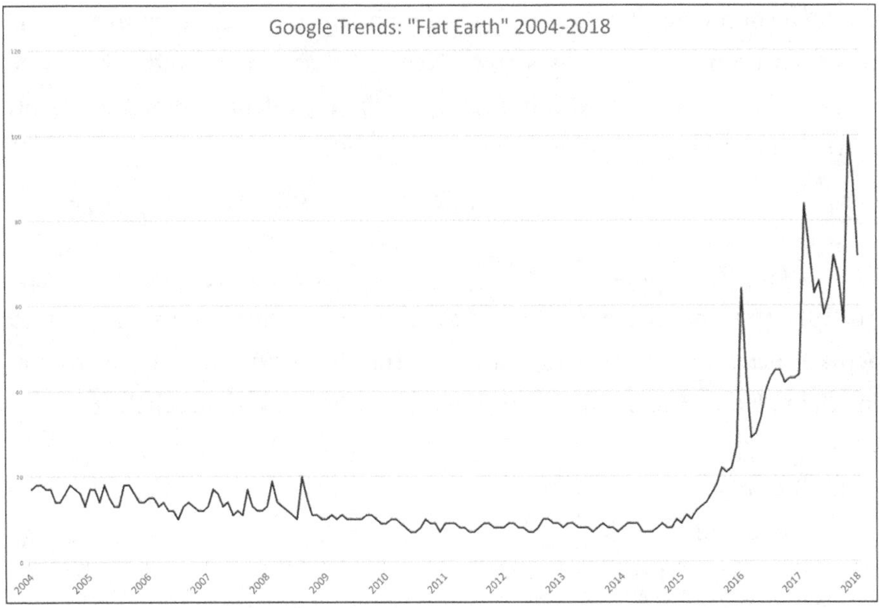

FIGURE 14: Interest in "Flat Earth" had been declining until 2015.

After the death of Johnson, Flat Earth existed on the internet at a slow burn. His organization found new life in the form of The Flat Earth Society, run by Daniel Shenton. There was a website and a reasonably popular forum,[2] but it remained largely unknown to the general public. When people stumbled across it they generally assumed it was just a big joke.

Then in 2015, its popularity began to rise. It's not clear exactly why, but there seemed to be a critical mass of YouTube videos. Several people were making reasonably high-quality videos repeating the exact same claims made in the 1800s. Now perhaps there was enough of them, allowing people to get sucked into this particular rabbit hole more easily.

The growth in Flat Earth interest has been meteoric over the last three years, eclipsing "Chemtrails" in popularity. Much of this is obviously a fad on the part of non-believers—just fascination with the "crazy Flat Earthers." But there are also many people who seem genuinely convinced that the Earth is actually flat and most of them got sucked in via YouTube.

It really hit the mainstream in January of 2016. Like any sufficiently interesting rabbit hole it had attracted its share of celebrities, like rapper B.o.B and reality TV star Tila Tequila, who both spoke out about their suspicions about the shape of the Earth. This was picked up by the entertainment media, and this then prompted responses from Neil deGrasse Tyson and others who explained that the Earth was in fact round. These celebrity rebuttals simply increased interest in the topics, and now here we are.

The Flat Earth community exists largely on YouTube and there's several YouTube personalities who make it their focus. It's not always clear if they actually believe their theories, but they certainly spend a lot of time promoting them. Here's the top five ranked by subscribers (as of March 2018):

128K Rob Skiba
94K Jeranism
66K Celebrate Truth
52K Mark Sargent
42K Mr Thrive and Survive

These numbers are really not very large in terms of actual YouTube celebrities who count their subscribers in the millions, but they are on the same level as the main Chemtrail channels like Geoengineering Watch's Dane Wigington (64K subscribers) or the biggest 9/11 conspiracy channel, Architects and Engineers for 9/11 Truth (43K subscribers). The Flat Earthers have the others beat on views though, Rob Skiba has 16 million views since 2013, Jeranism also has 15 million since 2015, but AE911Truth only comes in at 7 million since 2008. Dane Wigington does better at 7.5 million since 2014.

These numbers are important because they help bring perspective to people down the rabbit hole. Your friend might think that the Flat Earth conspiracy theory is very important, probably the biggest issue of the day, something that lots of people are interested in.

But it's not. There's literally tens of thousands of YouTube channels that get more views and more subscribers. There's channels that are devoted to odd things like dropping a red hot ball onto different things (carsandwater 835K subs) or crushing things with a hydraulic press (Hydraulic Press Channel 1.8 Million subs) or playing with magnets (Magnetic Games 288K subs).

One single video of a red hot ball dropped onto floral foam has had more views than Skiba's entire channel over its entire lifetime.

The modern Flat Earth movement's videos generally comprise of two things, either a long list of "proofs" that the Earth is flat, or a more detailed discussion of one individual claim. From a debunker's perspective it's very tempting to leap right in and start explaining whatever you first happen to come across. Flat Earth debunking is a rabbit hole in itself. There's hundreds, if not thousands of pieces of "proof" that supposedly show the Earth is flat. They are all based on fairly simple misunderstandings of geometry and optics, so are relatively straightforward to refute by correcting the mistakes. It's also fun, in a nerdy kind of way. The geometry and optics are relatively simple, and there are interesting problems to work through like "how far away is the horizon . . . ?" The problem is that most people don't understand geometry and optics, and even if you can explain one item to them, there's still hundreds more.

The best strategy is to focus on a handful of core beliefs, and address those in detail by showing undeniable proof that the evidence presented to support those beliefs is either wrong or is actually only explained by the round Earth. It's the same strategy as with dealing with believers in Chemtrails, focus your efforts on the core beliefs and resist the temptation to refute all their arguments at once. Flat Earth believers are fundamentally opposed to any form of argument from authority, so you are going to have to *show* them.

Over the Horizon

Flat Earth claims have changed very little since the 1800s. Modern Flat Earthers are simply repeating claims that were both made and debunked well over a hundred years ago. For example, MMA instructor and YouTube Flat-Earth celebrity Eddie Bravo said in November 2017:

> *The first proof [given] that we live on a ball is when ships sail off in the ocean they disappear over the horizon, hull first and then mast, so it appears to be going over a curve. So if something goes over a curve . . . would you be able to zoom in on it with a zoom lens?*[3]

Bravo thinks that if a ship appears to vanish over the horizon, if you were to zoom in on the horizon then the ship would reappear. This is a foundational belief of Flat Earthers, and one of the first things that you will have to address. It dates back to Rowbotham in 1865:

> A ship at sea, who outward bound . . . the hull disappearing before the mast, could be seen again . . . by a telescope, if the power were sufficient to magnify at that distance. . . .
>
> . . . if a vessel is observed until it is just "hull down," a powerful telescope turned upon it will restore the hull to sight.[4]

Notice how Bravo's description of the hull followed by the mast vanishing, then reappearing upon viewing with a zoom lens is very similar to Rowbotham's. It's likely because modern Flat Earthers are simply copying Rowbotham's playbook. For example, Eric Dubay in *The Flat Earth Conspiracy*:

> If you watch a ship sailing away into the horizon with the naked eye until its hull has completely disappeared from view under the supposed "curvature of the Earth," then look through a telescope, you will notice the entire ship quickly zooms back into view, hull and all, proving that the disappearance was caused by the Law of Perspective, and not by a wall of curved water![5]

The fundamental problem with this claim is that it's simply not true. YouTube videos that claim to show this effect invariably show ships that are simply too far away to observe details with the naked eye. Zooming in on the image makes them bigger and clearer, yes, but it does not restore any portion that was hidden by the horizon. This is something you can verify yourself with a sufficiently powerful zoom, and something that is a powerful practical demonstration for your friend.

FIGURE 15: This boat is partly hidden behind the curve of the horizon. Zooming in does not change this.

Figure 15 above shows a boat a few miles in the distance. The inset view shows that the hull is hidden by the ocean horizon. The ocean horizon is flat and unmoving, so that tells us it's not simply a wave (I watched the boat stay this same way for several minutes). The important fact here is that this a "powerful" telescope (2000mm, 83x zoom), capable of seeing the details of the boat, and the people on it. The boat was just visible to the naked eye, and looked the same as it did when zoomed in. At no point did zooming change anything other than apparent size.

Samuel Rowbotham was wrong in 1865, Eric Dubay was wrong in 2014, and Eddie Bravo was wrong in 2017. No doubt people will continue to be wrong in the future. But practical demonstrations like this are a powerful way of explaining things to people.

Viewing a small boat can be a bit tricky—a more practical, and less error prone, observation can be made of what is behind the boat, the distant mountain. The ideal setup for this demonstration is a mountainous island or headland about thirty to forty miles away, and a viewpoint on a beach, and some nearby place higher than the beach. The coast of Southern California is ideal for this, in particular Catalina Island is visible from beaches accessible to tens of millions of people. The procedure is quite simple: you take a photo of Catalina from the beach, then from a point higher up (like twenty feet above sea level), then again from the highest nearby point, like the top of the cliff in Santa Monica (or ideally from the top of a tall building).

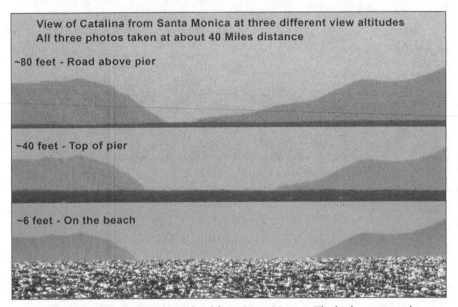

FIGURE 16: Three views of Catalina Island from Santa Monica. The higher you go, the more you can see. Exactly as you would expect on the globe.

The results are plain to see. When viewed from a cliff above Santa Monica Beach, the two halves of Catalina appear nearly touching (the island is just eighty feet above sea level in the center, with a mountain on either side). When viewed from the beach, the bottom third of the island is now missing, with a huge gap between the two sides. The middle view from halfway up the road shows something in-between the other two. Since we lose sight of a few hundred feet of Catalina simply walking down to the beach, this means it's behind something. The only thing between Santa Monica beach and Catalina is the ocean. So Catalina is behind the ocean, which means that the ocean, and hence the planet, is curved.

The great thing about looking at a mountain instead of a boat is that there's no confusion about if the visibility has to do with it being small or over the horizon. It also avoids ambiguity surrounding whether a boat has moved positions due to waves, as mountains tend to be rather fixed in position and elevation. You can see Catalina just the same (weather permitting) with your naked eye; even a cell phone camera can take images that demonstrate this effect.

The challenge here, should you happen to have a Flat Earth friend, is getting them to actually look at the evidence. Hopefully the allure of a trip to the coast might be temping enough. Take a zoom camera, take some binoculars, go to the beach, and look at islands partly hidden by the curve of the Earth.

Where Is the Curve?

A common refrain of the Flat Earther is "where is the curve?" If we live on a ball, they say, should we not be able to see the horizon curving from left to right? The simple answer is "no," as the ball we live on is very large compared to the amount we can actually see of it. If you fly up into space you'll see the curve, but at very low altitudes it's too gradual to see with the naked eye, and at higher altitudes it's generally obscured by clouds and atmospheric haze.

However, you can see the curve of things that go *over* the horizon. Lance Caraccioli, who posts on YouTube under the name "Soundly," has made it his mission over the last few years to collect clear and irrefutable images of this curve.

FIGURE 17: Pontchartrain Causeway demonstrated the curve of the Earth. Photo by Lance Caraccioli, a.k.a. Soundly.

Figure 17 above is one of the finest examples. It's the causeway over Lake Pontchartrain, just to the north of New Orleans in Louisiana. The causeway (basically a very long bridge) is a straight line with no slope. Because of this, and because it's a constant height above the surface of the water, it traces out the curve of the Earth. Ignore the two humps in the middle and follow the left edge of the causeway. Notice it goes up, over, and down. It's showing the curve of the Earth.

You can show this image to your friend, but it's a tricky shot to get in real life. The reason the humps in the middle look so dramatic is because of the extreme perspective foreshortening. You need to find a miles-long perfectly straight causeway, position yourself at a shallow angle to it, and then use a powerful zoom lens. If you can do that it would be ideal. But if you can't then have them look at Soundly's channel, where he's not only posted numerous different examples, but has also documented all the steps of the process he took to create the images.

The Size of the Sun

In the real universe the Earth is a sphere and it orbits the Sun, which is also a sphere. The Sun is really far away and really large, so even though you might be viewing it from places on the Earth that are thousands of miles apart (say, London and New York) the Sun appears exactly the same size in the sky. You can verify this by taking a photo of the Sun at sunrise, noon, and sunset. Roughly speaking when the Sun is overhead for you it's setting for someone else a quarter of the way around the globe to the east, and rising a quarter of the way round to the west. Noon in New York is sunset in Europe, and sunrise in Hawaii.

The problem for your Flat Earth friend, and the great opportunity for explanation, is that this would be impossible on a Flat Earth. Their idea of the world looks something like this:

206

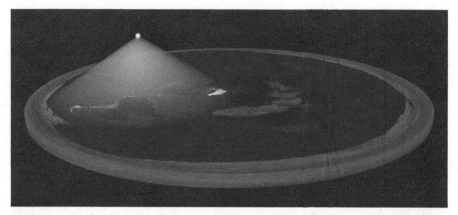

FIGURE 18: On the Flat Earth, the Sun works like a spotlight. This does not correspond to what we see.

In this model, the Earth is flat, and the Sun goes around in a circle above it. It uses some kind of "spotlight" effect to only illuminate the region below it, hence creating day, night, and time zones. The edge of the Earth is a ring of ice (the "ice wall") and nobody knows if beyond that there's empty space, an infinite plane of ice, or more flat worlds just a few thousand miles away.

This interpretation obviously has *numerous* problems. But let's focus on two that should be the easiest to wrap your head around, and that are very easy to verify with observation.

Firstly, if you look at the Flat Earth model above, there's nowhere on the surface of the Earth from which you can't see the Sun. South Africa is shown in darkness, and yet if someone were to be standing there then there's nothing between the Sun and that person. They should see the Sun, and it should still be quite high in the sky.

This is usually explained as some kind of strange function of perspective, however all perspective does is make far away things look smaller.

This brings us directly to the next point: the Sun does not vary in apparent size through the day. Consider someone standing below the Sun at noon on the equator, that means it's 3,000 miles above them (estimates vary but make no significant difference to the problem). Now if someone is 4,000 miles away on the surface of this Flat Earth, towards the edge of the cone of sunlight, that makes the actual straight-line distance to the Sun 5,000 miles.[6]

This means that through the day the Sun (on a Flat Earth) is between 3,000 and 5,000 miles distant. A general law of perspective is that if something is twice as far away, then it appears to be half the size. You can verify this yourself by taking a photo of something like a car from ten paces away, then from twenty paces away. This law does not change as things get bigger and the distance increases. Try it with a house at one mile and two miles, or a mountain at ten miles and twenty miles.

But since the Sun (and the Moon for that matter) stay exactly the same apparent size all day long, then that means it's not getting significantly closer or further away. Since it's moving across the Earth then that means its distance must be many multiples of the size of the Earth itself—so it must be millions of miles away. Therefore the Earth cannot be flat.

This can be difficult to communicate. Most people are not really that comfortable with basic geometry. Try to keep it as simple as possible. It's best done by pointing to the Flat Earth map. Show that the Sun would be closer at noon, and then far away at sunset. Get them to verify for themselves that it does not change size—I explain a few techniques on Metabunk.[7] As always, a practical demonstration beats a verbal explanation.

Ground Truth of the Stars

Perhaps one small reason for the ease with which some people get sucked down the Flat Earth rabbit hole is the decline of popular astronomy because of light pollution. Ancient asronomers made detailed observations of the Sun, the Moon, the stars, and the planets thousands of years ago, and it was very apparent to anyone with a dark-enough night sky that the Earth appears to sit in the middle of a sphere of stars that rotates around it. This sphere (called the celestial sphere) appears to be very far away. The relative positions of the stars in the constellations do not noticeably change, even if you move thousands of miles away.

The motion of the Earth makes the celestial sphere of stars appear to rotate about the same axis as the Earth. You can see this in time-lapse photos if you point your camera to the North. The stars all trace out circles around a point close to the North Star, which we call the northern celestial pole. This observation works out just fine on the globe Earth, and you can even roughly shoehorn it into a Flat Earth model by imagining it as a dome of stars rotating over the Earth.

Where the Flat Earth model doesn't hold to scrutiny is in the southern hemisphere. In the real world the southern hemisphere (Australia, South America, Southern Africa, Antarctica) behaves as a mirror image of the northern hemisphere. The stars appear to rotate in circles around the southern celestial pole.

But how would this work on a Flat Earth? If the stars were on a dome, then the stars in the South, when viewed from Australia, would not appear to rotate around a southern point, but instead would be whizzing by in nearly straight lines at very high speeds.

As always, in an ideal situation you would get your friend to observe this for himself. Taking a time-lapse of the stars circling the northern celestial pole is relatively easy. You can even do it with a phone app like NightCap which has modes to automatically record the paths of the stars. The northern celestial pole can be easily photographed from North America. The southern pole can still be detected, even though it's well below the southern horizon, you can still point your camera south and see the star trails as they arch up and over the celestial pole—totally unlike what they would do if they were on a dome.

There's an easier way than going out into the cold and fiddling in the dark with your camera. You can simply use a night sky simulator app to get the same view from the warmth of indoors. My favorite is Stellarium, which is both free and very comprehensive. It allows you to set up a view from anywhere on the Earth at any time and see what the sky looks like from there. You can also speed up time, so you can position the camera in Australia, look south to see the stars rotate around the southern celestial pole, and look north to see the trails arc up over the northern celestial pole.

The objection that will be raised here is, "Why would you trust a computer program over your own eyes?" The answer is that *Stellarium has never been wrong*. Millions of people use it all around the world. None of them have even reported that the image of the sky that Stellarium generates is any different to what they actually see.

This is a concept known as "ground truth"—the verification of distant observations or models by actual observations on the ground. Stellarium has a lot of ground truth, so you can use it with confidence that what you are seeing is what you would actually see if you were to go to that particular place and time yourself.

It's not just Stellarium either. There are many other programs, and many easy phone apps like Pocket Universe that allow you to point your camera up at the sky and they tell you what they see (allowing you to form your own ground truth). These apps also allow you to change your time and location with a few taps.

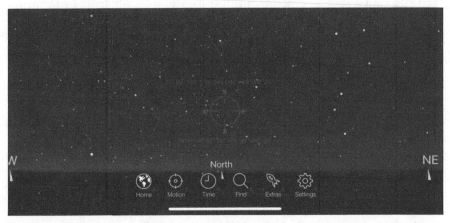

FIGURE 19: Typical iPhone app "Pocket Universe," with which you can prove, again, that the Earth is round.

Have your friend travel virtually to Australia and look at the stars rotating north and south, repeat for South Africa, and Chile. Compare with similar points in the North. This symmetry of rotation can only work if we are on a ball in the middle of the celestial sphere.

It's easy to see how people might have been sucked into the Flat Earth rabbit hole back in the 1880s. The world was more local. There was no aviation, let alone space flight. Communication was slow and claims about the shape of the Earth were hard to verify. Today there are photos of the Earth from space, cameras that can see ships drop below the horizon in high resolution, instant communication with friends in the southern hemisphere, and a wealth of verifiable knowledge. There are no longer any reasonable excuses for believing the world is flat.

Flat Earth Summary and Resources

Key Points to Convey to Your Friend
- The modern Flat Earth theory is only around 150 years old. We've known the Earth was round for over 2,000 years.
- Most Flat Earth points are just repeating what was written and debunked in the 1800s.
- You can go to the beach and see things like boats and islands vanish behind curve. Zooming in does not bring them back.
- If you find a long straight causeway or line of pylons you can see them curve up and over the horizon.
- The Sun does not visibly change size through the day. It also sets below the horizon. Both are impossible on a Flat Earth.
- The stars form a "celestial sphere" around the Earth that acts the same in Australia as it does in the US. On a Flat Earth it would be totally different.

Additional Resources
- Metabunk's Flat Earth forum—**metabunk.org/forums/Flat-Earth/**
- Popular Science—10 easy ways you can tell for yourself that the Earth is not flat. **popsci.com/10-ways-you-can-prove-earth-is-round**
- David Ridlen's Flat Earth Debunk Visualization—What a Flat Earth would *actually* look like. **youtu.be/uexZbunD7Jg**
- HD Images of Earth from Space—The entire planet in HD every fifteen minutes, looking like a ball. **metabunk.org/.t8676/**
- Mirage Simulator—A useful tool for demonstrating why you can see things across a lake that are normally hidden. **metabunk.org/mirage/**

CHAPTER FOURTEEN

Bob – Escape from Flat Earth

Bob is a young man from Canada who once thought the Earth was flat. He was in his mid-teens when he first became interested in the idea. Like many young people emerging into young adulthood, he was simultaneously searching for meaning in the universe and feeling the potential strength of his own intelligence in determining what that meaning might actually be.

Like most of us at that age, he somewhat overestimated his abilities.

I was in a certain state of mind where I wanted to separate from society. I wanted to feel special, so I put a lot of effort and interest into science and philosophy. Unfortunately my judgment was weak and so I started to believe in and "join" a lot of causes. I discovered Flat Earth mainly through YouTube. It felt good. I felt like this was some kind of purpose for me, and that I was unique. After I joined the Flat Earth movement I felt like I was more important than everyone else.

I actually thought that the Earth was flat. I thought there was nothing true about the "Globe Earth" idea. I believed that the government was always lying about everything—although I still believe that to some degree, with better judgment. Back then though I was like a rebel. It was a silly time in my life.

Like most people in recent years, Bob was sucked down the rabbit hole via YouTube. Like Richard, this happened when he was still young, still searching for new ways to view the world.

The YouTube videos that had the most effect on me were when they said: "you can't see the curvature by yourself" and "because they teach us from a young age to believe the Earth is round even though an individual can't know for sure." I accepted the idea that the Earth was flat because I rejected all the conventional logic from the Globe Earth perspective. I thought it was all lies because I thought people who believed the Earth was round were silly and closed-minded. Now I realize it's more like the opposite.

Bob was not so much swayed by the accuracy of the Flat Earth evidence, or by the logic in the arguments. Instead he simply assumed that conventional thinkers were the closed-minded ones. He was afire with a sense of enlightenment, of "waking up," and the actual sober facts were not going to get in his way. Very much like Willie with the Apollonian (logical) and Dionysian (instinctive) thinking, Bob had put down roots on one side of the fence—the side that *felt* better.

This decision didn't come from nothing. Bob actually knew some people who seemed to genuinely believe in the Flat Earth theory—and not his teenage peers, they were, in fact, quite the opposite.

Yes, I know people who believe in Flat Earth. These are generally older people or religious people. There's three of my elderly relatives who are believers. They believed in Jesus, God, the Bible, and also Flat Earth. They never tried to convert me, but we did have some discussions after I became a believer.

Flat Earth is one of the few conspiracy theories where religion can have an incredibly strong effect. The rejection of science required to believe in the Flat Earth is so great that it requires something similar to a deep religious faith or dogma. Many of the older Flat Earth believers you might come across have a strong religious nature. In Bob's case the effect was indirect as he was not particularly religious. Although his elderly relatives did not seek to convert him directly, the fact that there were people he grew up with who were already down there probably eased the transition into the rabbit hole. Without this foundation, the YouTube videos might never have gained the purchase that they did.

Bob did not discuss his ideas with other members of his family, but he did discuss them with friends.

I hid it from my family because I knew they were going to judge me a lot. I did talk about it with my friends. At first they thought I was joking, but then I kept arguing. They tried to show me I was wrong, but for every piece of evidence they brought up I just thought that it was fake, and part of the conspiracy. I thought my friends were not smart enough to understand that they were wrong. I was stubborn, and I refused to listen to them.

Here there are two groups who were in a position to help him. Unfortunately, he feared judgment from his family and so he never got to explore his ideas with them. His friends were able to talk to him, and even tried to bring some evidence and logic to the discussion, but he flat-out rejected them, regardless of what they said. Eventually he realized what he was doing.

I began to see that the Flat Earth movement had no real argument. They brought nothing new to the debate, they just dismissed all the Globe Earth arguments just by saying that it might false, or it might be part of the conspiracy, and "we can't know for sure."

Someone told me that some people are like irrational jealous wives, they think you are cheating on them, and if you say you were at a bar with a friend then they say that's exactly the excuse a cheating husband would use. The thinking just gets crazier with bigger reasons not to believe someone.

I realized the crazy wife in that story was like the Flat Earth people. This made me think I should listen to the Globe Earth point of view, and eventually I began so see that it made more sense.

Bob's turning point seems not to have been one particular piece of evidence. Rather, prompted by his friend's story, it was a realization that the way he had been thinking and arguing was really not as smart as he thought it was. Conventional arguments about the evidence did not seem to work with him.

People on the internet who tried to convince me were wasting their time. I just ignore arguments and evidence that came from normal people like me. It just did not go anywhere.

But I watched this one video about conspiracies related to space exploration. It was not a debunking video, but it was explaining why those ideas are attractive to so many people. I was lucky that I was not so single-minded that I only watched videos I believed in. Without influences like this video I would probably still be a very closed-minded Flat Earth believer. I watched some other content like this and it slowly triggered doubt, enough to help me come back to reality.

Now Bob is out of that rabbit hole and casts a more critical eye upon other conspiracy theories.

I still look at some conspiracy theories that seem credible. I don't believe in Chemtrails. With 9/11 I think maybe Bush did it, maybe not. I can't be sure because there's not enough evidence. But with theories where there is evidence I can have an idea. Like the Moon landing, we did land on the Moon. There's obviously no lizard people ruling the world.

I sometimes try to help other people. But I honestly think some people can't get out of their comfort zone and they will never change. They don't look for truth, they look for something easy for them to accept or that makes them feel special. I don't have a lot of hope for the Flat Earthers. Once I turned back to being a globe believer, I just avoid the subject with my elderly relatives. I know they have lived all their lives with this idea, so I won't be able to make them realize their mistakes. If I joined a Flat Earth Facebook group, it's more for a good laugh. If I see a comment from a Flat Earth believer I will leave another comment to maybe make them use their brain and logical thinking.

Don't mock them. That will worsen the situation, it will make the person feel even more isolated and have trust issues. When my friends thought I was joking about believing in Flat Earth it didn't help me see their point of view. It just made me want to get rid of them and find other friends who would understand me.

But if you want to help someone who is down the rabbit hole you really should make them feel like you are not putting them down for what they believe. Make them feel that you are open to what they say, even if you know they are absolutely wrong. You might open a door to reason for them.

The Flat Earth theory is at the far end of the conspiracy spectrum and this extremeness has an effect on how people get in and how they get out. Bob's story shows how important it is to give it time. He initially rejected all arguments against his position without even considering them. What got him to take another look was gaining some perspective on his own thinking and how it relates to how other people think. This took time because he had to establish a pattern in his own actions before he could recognize it as a pattern. To some degree he had to get deep enough inside the rabbit hole to realize he was in one.

That he did not get stuck is in part due to the resilience and adaptability of his young mind. But it is also due to him having friends who were willing to talk to him without fully alienating him. This aspect was far from perfect. Being laughed at by some of his friends nearly made him want to fully separate from normal society. But in the end it was a friend who tipped the balance, someone who told him the story of the overly suspicious wife. The story clicked at the right time, as he was starting to see the flaws in the Flat Earth camp. It became the start of his escape from the rabbit hole.

Bob's story demonstrates some of the complications of debunking. The various proofs and arguments around the Flat Earth theory quickly get quite technical, with geometry, sines and cosines, square roots, etc. The average person can quickly find this aspect impossible to grasp, and so they, like Bob, start to judge arguments on a simpler basis, like how they feel, or who made them.

Another complication is that of family. Bob's elderly relatives helped get him into Flat Earth, but once he got out he recognized there was little to be gained, and something to lose, by trying to convert them to the globe. His more immediate family played another role, in that he was afraid to discuss his beliefs with them for fear of mockery. Family relationships require special consideration.

I'll discuss these complications, and others, in the next chapter.

PART THREE

CHAPTER FIFTEEN

Complications in Debunking

In 2017 Charlie Sheen starred in a movie called *9/11*. Sheen had previously been associated with 9/11 Trutherism and was asked about this by the *Hollywood Reporter*. He responded:

> *I know I got lot of heat for the opinions I had that weren't just my own, I was not just coming up with stuff about 9/11. I was parroting those a lot smarter and a lot more experienced than myself, who had very similar questions. . . . I am more about moving forward, . . . we must never forget, but there are still a couple of things just rooted in simple physics that beg some measure of inquiry.*[1]

"Simple Physics" is an oxymoron. There is no simple physics. If it seems simple that means you are reducing something that is complex. The great physicist Richard Feynman was once asked by an interviewer about magnets: How do they work? His answer was basically, "I could tell you, but you wouldn't understand." He then tried to explain *why* the interviewer would not understand:

> *I can't explain that attraction in terms of anything else that's familiar to you. For example, if we said the magnets attract like rubber bands, I would be cheating you. Because they're not connected by rubber bands. I'd soon be in trouble. And secondly, if you were curious enough, you'd ask me why rubber bands tend to pull back together again, and I would end up explaining that in terms of electrical forces, which are the very things that I'm trying to use the rubber bands to explain. So I have cheated very badly, you see. So I am not going to be able*

to give you an answer to why magnets attract each other except to tell you that they do.[2]

Feynman here is asking you to trust him. He's telling you that's really your only option, because to understand the real physics behind magnetism you'd at least need to take a few undergraduate courses, and quite possibly more than a few. There is a more complicated answer, but it's not currently accessible to you without a lot of work.

But some people demand an immediate explanation without doing that work. Why indeed did the World Trade Center towers collapse the way they did? They may cite Newton's laws of motion and tell you that the collapses violated them unless explosives had been used. They ask you to explain that, but unfortunately, like many perfectly reasonable and intelligent people, they find themselves incapable of easily understanding the answer. You tell them that Newton's laws only apply to abstract point masses, and they will tell you that's bullshit. You try to tell them about point masses versus rigid bodies versus articulated bodies, elastic versus inelastic collisions, conservation of momentum versus conservation of energy, potential energy in the building versus chemical energy from explosives, static force versus dynamic force, vertical support cross-sections, and the square-cube scaling law. They tell you that's all a bunch of hand-waving because they think "every action has an equal and opposite reaction" proves the building could not have simply collapsed as they did.

Conversations like these are a challenge. The challenge is not a reflection on the intellect or education of your friend. Professor Feynman was not insulting the interviewer when he explained why the interviewer would not understand the full explanation, he was simply noting that it's more complicated than they thought, and you need a certain amount of education to understand it—or at least you need to put in a certain amount of time and effort.

The challenge is magnified if, like most people, you don't really understand it yourself. I'm a reasonably technical person when it comes to basic principles and calculations in physics, but like everyone I have my limits. I've forgotten most of the advanced math I learned at school, retaining only the strong practical subset that I used daily for years in my work programming 3D physics for video games. At some point when helping a friend understand something we are forced to accept the limits of our own understanding of why something does a certain thing, except to tell them that it does.

Charlie Sheen thinks there are questions raised by "simple physics," and yet he simultaneously defers to "those a lot smarter and a lot more experienced than [himself]" in order to ask those questions on his behalf. But if the questions are so simple, then why are there no simple answers? Sheen has just chosen to trust one small set of "smarter" people who argue that these "simple" questions are best answered by explosives, and ignore the other (much larger) set who recognize the question is somewhat more complicated, and that gravity, fire, mechanics, the physics of materials, and a rather complex sequence of events are required to provide the answer.

Show, Don't Tell

The best way of explaining something your friend does not really understand is to sidestep the entire explanation stage, and instead *demonstrate* that thing. For example, people who think the Earth is stationary might argue that if it were in motion, then when we jumped up in the air we would come down in a different place. Now you don't need to get into any great scientific detail here to debunk this. Discussing gravity, velocity vectors, inertia, and suchlike is not going to help. All you need to do is note that if you jump up and down on a train you'll land in the same spot. Tell them to try it out.

The "jumping and landing" problem is certainly not something that is intuitively obvious. Back when I was in the games industry a friend told me about his time working on a PlayStation game called *Blasto*, where at some point the player's character would ride around on little floating platforms. One of the game designers wanted it programmed so that when the player jumped straight up when on a moving platform, they would not land back on it.

That in itself would not be unreasonable, as video games have all kinds of weird physics, but the designer wanted it because he thought it made the game *more realistic*. My friend argued with him, and eventually settled on a bet. He would stand in the bed of a truck driven along the alleyway at a steady fifteen miles per hour, and then jump straight up. If he landed back in the truck in the same spot, then that is how they would program the game. If the truck moved away from underneath him and he fell to the ground, then they would program it that way.

Since he was moving at the same speed as the truck he landed back in it just as you would expect. The best way of getting the point across was a practical (albeit dangerous) demonstration.

Practical demonstrations are something that I often find myself doing and are one of the more enjoyable aspects of debunking. It's quite easy to simply keep on typing, trying to explain some mistake to someone. But if you can actually show it to them then that generally works a lot better. Stop trying to explain some aspect of physics with words, and instead spend that time concocting a good demonstration.

While they are fun and often effective, practical demonstrations need some caution. There are, after all, numerous videos online that claim to provide evidence *for* false conspiracy theories. This is especially true in the area of 9/11 controlled demolition. In one video, Richard Gage (head of Architects and Engineers for 9/11 Truth) drops one cardboard box on top of another, and then claims this demonstrates why the collapses of the World Trade Center Towers were so suspicious.[3] Many people were convinced by this, and have even repeated versions of the "experiment" themselves. So you need to make sure that the principles and assumptions underlying your demonstration are correct. It's a good idea to check with an expert before putting it out there.

Family Debunking

A few years ago, a relative emailed me about something that I immediately knew was a scam. She first told me that she was going to take advantage of her retirement to go into business for herself. What's more, she had already taken the first steps. It was a great new opportunity to get in on the ground floor selling natural products to her friends and family.

My relative, who I'll call Betty, had recently befriended a nice couple at some local event. After seeing Betty a few times they asked if she'd be interested in a way of making a bit of extra money. Politely Betty said yes, and they invited her round for dinner so they could discuss it. After dinner they brought out a case that was full of sample bottles of products like skin lotion, toothpaste, and shampoo. They told her that most of the products on the market had dangerous chemicals like sodium lauryl sulfate (SLS). They got a tube of Colgate toothpaste out and had her read the list of ingredients (which included SLS and other chemicals). They then showed her documents from the CDC that proved SLS was poison.

This was a bit concerning to Betty, who was generally in favor of natural things and opposed to poison. What could she do? Well the answer, they told her, was to use these natural products instead. They showed her the list of SLS-free ingredients. All at very reasonable prices. And what was more, she was in luck! It just so happened that they were selling more than they could handle and were looking for people who could sell the products for them, and look at this graph of how much money you can make, and even more if you could sign up other people to sell for you! They left her with the case of products to think it over.

This was just an old scam—multi-level marketing (MLM). MLM is a kind of pyramid scheme where the selling of a simple product is not the real money maker—the real money is actually in signing up people to sell products. The company and top salespeople make money selling these sample kits and getting signing fees, and hardly anyone else actually makes any money. Those who do make money often end up alienating those around them with persistent requests to sign up. It's an old scam, and it sucks people in.

I looked up the company, and sure enough there was a litany of complaints about it being an MLM scam. As always with these things there were a number of true believers who actually thought MLM was a good way of making a living. Statistics don't lie though, and according to the FTC over 95 percent of participants in MLM schemes end up losing money.[4]

What was I to do with Betty, someone who I was very fond of? I knew she was also an intelligent woman, but perhaps a bit more into "natural" remedies (like aromatherapy) than I was. I knew she liked this couple and did not want to offend them, and I did not want her to feel like I was criticizing her choice of business. I knew she respected me, and would probably listen to me, but I also knew that I had a limited opportunity to make my case and avoid her getting sucked in.

I took a very slow approach, starting by looking at the claims they had made about SLS being a poison. Indeed, it was, and there was a Material Safety Data Sheet explaining just how toxic it was. Then why was it in toothpaste? Like with all chemicals, it's the amount that makes something toxic, not the substance itself. Table salt, for example, is highly toxic—a tablespoon can kill a small child. But we still sprinkle it on our food.

This could be a bit of a tricky thing to get across to Betty, so I also had a look at the "natural" ingredients in the products she was planning to sell. Funnily enough those ingredients *also* had Material Safety Data Sheets explaining how toxic they were. Most usefully, some of the ingredients were actually listed as being *more* toxic than the supposedly dangerous SLS. I relayed all this to Betty the next day, without any judgment, just pointing out the facts. After it sank in, she said she'd return the sample case. "I'll tell them Mick made me do it," she said. We laughed.

She returned the sample case to the couple and they tried to talk her out of it. "What does this Mick know?" they asked. "He probably just read some disinformation on the internet. Just try it for a week." But ultimately Betty recognized that what I'd told her was probably correct, so she declined, and unfortunately their friendship was over.

Telling someone that they are wrong can be difficult on many levels. It can be even more difficult when you love that person and you know that your rejection of their belief might hurt your relationship, or their relationship with others. In such a situation the normal advice still holds: foster effective communication, supply them with useful information, and give it time. But the personal factor brings some complications.

First off, is it really worth it? Consider the pros and cons. Is their belief in conspiracy theories actually at a level where it's a problem—to themselves or others? Many people believe in strange and irrational things like the supernatural without it affecting them negatively. Do you want to risk creating conflict for something that's really not that important? In Betty's case the conspiracy was low level ("Big Pharma" supposedly covering up toxic ingredients), but the financial consequences of her believing it were very real and very negative.

If you decide it is worth it, then continue with caution. Your close relationship can be a benefit—in that you have a better understanding of what makes them tick. But it creates a problem in that your friend (spouse, lover, partner) expects that you will understand, accept, and *support* them. It becomes especially important to carry out discussions in a constructive and positive manner. Validating their genuine concerns and establishing common understanding initially takes precedence over pointing out where they are wrong and where their sources are wrong. Effective debunking isn't about scoring points.

When you do come to pointing out mistakes, do it (initially) only with the most neutral and inarguable facts, like the ingredient lists with the cosmetics. With Chemtrails you would avoid subjective arguments like "I think contrails have always been this persistent" and focus on neutral information, like the old books on clouds.

Finally look after yourself. Remember that your friend may look at your beliefs in the same way you look at theirs. They might be disappointed in you, they might get frustrated by you seemingly not taking them seriously, they might begin to attack and criticize you. Do not take this personally. Try to use it as a basis for conversation. If that does not seem to be working, then back off. Consider the importance of your relationship relative to the danger of a hasty response. When helping someone out of the rabbit hole is personal you have to be especially slow and careful. Unless you need to quickly prevent something like an unwise investment, then you should give it a lot of time. Recognize that it might be something you have to deal with for an extended period, but don't give up.

Morgellons

The word "Morgellons" means different things to different people. To those people who think they have it, then "Morgellons" is a disease with many symptoms, the most notable one being small fibers emerging from their skin. To the majority of the medical profession, "Morgellons" is just a name that some people give to their various medical problems when they are unhappy that the doctors are unable to find exactly what is wrong with them. When the patient is convinced that fibers are emerging from their skin, then the doctor might describe that aspect as "Delusional Parasitosis."

Morgellons is the most recent version of similar conditions that have been recorded for over 100 years. The modern term was coined in 2002 by Mary

Leitao, who was searching for answers for the skin conditions plaguing her son.[5] She did some internet research, copied and extended some symptom lists from the National Unidentified Skin Parasite Association, and the Morgellons Research Foundation was born.

The primary problem with Morgellons is that the patients are convinced there is something specific wrong with them, but their doctors are uncertain what it is and often suspect a degree of mental illness is involved. This creates conflict between patient and the doctor.

I spent a couple of years writing about Morgellons and interacting with Morgellons patients online. It's not exactly a conspiracy theory (although many believers will point a finger at "big Pharma" and there are some more extreme beliefs relating Morgellons to nano-machines and Chemtrails), but there are some similarities in the ways the believers think about the topic—particularly with confirmation bias. Every case is different, but there is some general advice I can give if you want to talk to a Morgellons sufferer.

Defer to doctors. You are (presumably) not a doctor, so do not give medical advice. Ask your friend what the doctor says. If they ask you what you think they should do, then suggest they talk to a doctor. If the doctor prescribes medication, then encourage them to take it as prescribed. Do this in a non-judgmental way.

Avoid discussing mental illness. The quickest way to have a Morgellons patient turn against you is to have them think you are suggesting they are mentally ill. While there are certainly mental issues involved with some patients, it does not help to discuss mental illness as a *cause* of the symptoms. You can discuss it in a supportive way as a *result* of the condition—itching causing sleeplessness and stress, for example.

Don't focus too much on fibers. Fibers are everywhere, so you will find them on your skin. You can explain this, but your friend is emotionally invested in the idea that the fibers are related to their illness. Instead of just dismissing them, you can suggest that at least some fibers are just clothing fibers, but you (honestly) don't know about all of them.

Discuss symptomatic, palliative care. If they are itching, then treat itching. If they are anxious, then treat anxiety. Accept that the cause of their problems is mysterious and try to shift the focus onto getting the most they can out of life by dealing with the symptoms. A "cure" would be ideal, but in the meantime let's work with what we have. There's no shame in treating the symptoms even if you dispute or don't know what is causing them.

Talk about other things. Many people who have self-diagnosed with Morgellons have some degree of hypochondria or "illness anxiety." Health issues have become an obsession, which probably translates to spending hours reading things on the internet. Encourage your friend to read about and engage in other things. Avoid enabling their obsession. If they continually discuss health with you, then try not to get drawn in; steer the conversation to other things.

Give it time. Like any problem involving strong beliefs, change can take a while. During that time, it might feel like you are not making progress. Remember it's a cumulative process, and change can happen slowly over time, or suddenly after months have passed. Improvements might be partial, or very minor.

While these are simply thing I've learned from talking to people who think they have Morgellons, they also tally well with common lists of advice given for hypochondria[6] and OCD.[7] You can read about those topics and apply much of what is written to your friend with Morgellons.

Mental Illness

Most people who believe in conspiracy theories are not mentally ill. Both my personal experience and the current scientific research indicate that conspiracy theorists are just regular people. As Cass Sunstein and Adrian Vermeule put it:

> *Conspiracy theories typically stem not from irrationality or mental illness of any kind but from a "crippled epistemology," in the form of a sharply limited number of (relevant) informational sources. Those who hold conspiracy theories do so because of what they read and hear.*[8]

Sunstein and Vermeule wrote that in 2008, and ten years later would probably extend that to "read, hear, and watch." The average conspiracy theorist is no more mentally ill than the average football fan or the average physics professor. It's a grave mistake to label someone as mentally ill just because they think the World Trade Center was destroyed with explosives. They are generally being ordinarily rational with the limited information they have.

But, as with Morgellons, actual mental illness can play a role in conspiracy thinking, both as a root cause, and as a complication. Some people do actually believe in conspiracies because of their mental illness, and some people have their mental illness made worse because of their beliefs in a conspiracy theory.

What generally distinguishes the genuinely paranoid conspiracist from the average conspiracist is the degree to which the conspiracy involves them personally. They might think that they are being followed or that their house is being searched because of their conspiracy activism. They might feel they are a victim of "gang stalking," where everyone around them is subtly trying to undermine them in some way. These types of delusional beliefs are known as "delusions of reference."

I have occasionally encountered people on the internet who are potentially mentally ill, and a few times in person. My general strategy when dealing with someone who I suspect is mentally ill is to briefly treat them as if they are not. Assume the best, and see what happens. But as soon it seems likely that they are in fact mentally ill then I will not continue to engage them about conspiracy theories.

I am not a doctor, and I am not a psychiatrist. I'm a debunker, a fact-checker, and a communicator. I do not know how to cure illness, mental or physical. My attempts could quite possibly make the situation worse. If I have debated you on the internet, then take comfort in the fact that I do not think you are mentally ill.

But what if your friend exhibits signs of mental illness, or even if they have been diagnosed with some mental condition? If you cannot walk away, then how should you deal with the role of conspiracy theories in your friend's situation?

It will vary, but probably your best approach is going to be to avoid discussing conspiracy theories. The problem here is not some misunderstanding of physics or chemistry, or even a "crippled epistemology" that you can repair with large amounts of correct information. The problem here is mental illness. The conspiracy theories might be a result of that, or they might just be making it worse. But discussing something they are obsessed about is probably not going to help your friend.

Instead, be a good friend. Communicate with them about more neutral matters. Encourage them to listen to their doctors. Steer them away from conspiracies. Give it time. If you want to do more, then seek advice from a mental health professional.

Political Debunking

The topic of politics is notoriously fractious, a minefield subject around the Thanksgiving table. It is often recommended that for the sake of family harmony we avoid discussing politics (and religion) at all.

The arguments that people get into over politics are, to a significant extent, low level conspiracy theories. Conservatives think that liberals are conspiring to tax them into oblivion, to take away their guns and property rights, and to rig elections by legitimizing illegal immigration. Liberals think conservative are conspiring to make the rich richer, restrict the voting ability of minorities, collude with Russia, and silence environmental scientists. Both sides think the other is spreading misinformation and conspiracy theories.

The techniques outlined in Chapter 5 have direct relevance here, with some modification. Effective communication, finding common ground, being respectful, validating genuine concerns, and supplying missing information are all great things to strive for. But political discussions have some unique aspects that make them differ from conspiracy theories like Chemtrails.

Firstly, there's the symmetry of perception. Much like with 9/11, your friend will quite often know a lot of individual facts related to the topic, quite possibly more than you. They probably have a strong belief that you are the person who is misled, and that it's their job to explain things to you.

This symmetry of perception is usually matched by an asymmetry of knowledge. You will have one idea about the facts of the matter, they will have another. Both of you will generally have significant gaps in knowledge, and specifically you will have gaps in understanding what the other person believes and why. Closing those gaps is a necessary step towards having a discussion.

There are many extreme political conspiracy theories, with varying degrees of plausibility, things like "Pizzagate," "The Storm," "QAnon," and "Russiagate." But you are unlikely to make much headway by diving into such convoluted examples as these. Before tackling something like that, try to focus on something where you can at least look at the facts behind the theory.

For example, in November 2015, Donald Trump (then the Republican presidential candidate) appeared to mock the disability of reporter Serge Kovaleski. Trump waved his arms around with his hands at an odd angle and said, "Now, the poor guy, you've got to see this guy: 'Uhh, I don't know what I said. Uhh, I don't remember.'" Kovaleski suffers from arthrogryposis, a condition which restricts the positions of his hands, and Trump did appear to be mimicking and mocking Kovaleski's appearance.

To many people looking at the footage of Trump side by side with footage of Kovaleski, this seems like an obvious conclusion. Yet Trump supporters will tell you that it's a conspiracy, that's it's fake news spread by the liberal media, and that the accusation has "been debunked."

How is this possible? And how can the conversation continue after such polar opposite interpretations have been revealed? Often the discussion will immediately degenerate into anger because the liberal will be disgusted that their conservative friend is letting Trump get away with mocking disability, and the conservative will be disgusted that their liberal friend has blindly followed fake news.

In debunking conspiracy theories I've recommended that you first work to establish common ground before moving on to supply missing information. In political discussion these two steps need to be combined into one: mutual sharing of information.

To achieve common ground, you need to understand how the other person has formed their opinion, and they need to understand how you formed yours. With Kovaleski, conservatives feel that the accusation of mocking disability has been debunked because Trump had previously used exactly the same hand motions when mocking his then opponent Ted Cruz, who suffers from no physical disabilities.

Trump opponents might be reluctant to concede this. But the point is not to excuse Trump's actions, but rather to understand why Trump supporters believe the way they do about the situation. This is missing information on the part of the Trump opponent, who often will simply outright reject claims like "that has been debunked" with disgust, not looking into it because they feel the mocking of a disabled reporter has been proven beyond any doubt.

But both Snopes and PolitiFact (supposed bastions of liberal bias) take a more nuanced view of the situation, with Snopes saying it was "a subject of debate" and PolitiFact saying Trump's actions were "mocking, whether a disability was involved or not." This might be missing information to the Trump supporter, as might be the information that Trump and Kovaleski were on first name terms for years, and Kovaleski is certain that Trump would remember his condition.

Unencumbered truth is a rare occurrence in politics. The goal of the rhetoric of politicians and their supporters is to sway public opinion in order to win elections and gain or keep power. Regardless of which side you are on, it is unlikely you have a clear picture of the facts surrounding a topic if you only get your information from sources that you feel are on the right side of your ideology.

At the very least, by restricting your information sources you will be missing out on the spin that has caused your friends to believe what they do. Not having this information will prevent you from finding common ground, which is crucial to moving the conversation forward in a productive manner.

The common concession in debates that "the truth is somewhere in the middle" is a fallacy. The world is not half flat, jet planes are not half spraying poison, the World Trade Center was not half demolished with explosives. But politics differs from these traditional conspiracy theories in that it's an appeal to the center and not an appeal to an extreme.

Both sides seek to exploit every situation. A *Bloomberg* poll in August 2016 found that likely voters thought that the mocking of Kovaleski was the worst thing that Trump had done. Trump opponents were not going to let a little nuance get in the way of this valuable tool.

Similarly, the notion of "fake news" was powerful rhetoric that resonated with Trump supporters. The more the liberal media talked about the

Kovaleski incident, the more the conservative media could point out they had "debunked" it, and hence support the idea of liberal media bias.

Common ground is not going to be immediately found in the middle. The first piece of common ground to be found is when both you and your friend explain to each other why you hold the position you do, and when you supply each other with information you think is missing. You don't even have to agree with the validity of the information—just acknowledge that it exists, and that it's a foundation of your friend's beliefs.

From there you can get into the nitty-gritty of fact checking. But you might want to take a step back and refocus the discussion on more important topics. Once you both agree upon the basic facts then topics like the Kovaleski case might not be worth pursuing.

Instead use the discussion to encourage your friend to seek out other information sources on other topics and make a commitment to do the same yourself. Look at the spin coming from the other side, but also look at more neutral sources, like people who share your own ideology but are now largely retired from public life. Look to people who might once have been considered political party heroes, like Ronald Reagan or JFK—what did they think on the topic? What has changed?

It might seem like an impossible task, but be polite and respectful. Politics is polarizing because politicians want it to be polarizing. They don't want there to be any possibility that their followers would get drawn over the line to the other side. To have a fruitful political discussion with your relatives over Thanksgiving you need to sidestep the polarization, the spin, and the conspiracy theories. Instead focus on *mutual* understanding, focus on facts, and focus on the true differences between you, and the true similarities.

CHAPTER SIXTEEN

The Future of Bunk and Debunking

Jenna Abrams was a popular figure on Twitter during the 2016 presidential campaign, amassing over 70,000 followers.[1] Abrams, tweeting as @Jenn_Abrams, started out in 2014 with a constant stream of tweets that reflected the populist right-wing politics of a demographic that would end up being a core part of Donald Trump's base. Her more popular tweets were increasingly retweeted by establishment figures like Donald Trump Jr. and Kellyanne Conway (then Trump's campaign manager).

The problem is Jenna Abrams never really existed. According to congressional investigators the account was in fact the creation of a Russian government-backed entity called the Internet Research Agency. Based in St. Petersburg, Russia, this agency employed hundreds of people with the goal of spreading information and misinformation that would undermine the US, and promote the interests of Russia.

These workers created thousands of what are known as "gray outlets" or "trolls," social media accounts and web pages that look at first glance like the pages of ordinary Americans and other Westerners but were in fact being run by English-speaking Russians, or Westerners in the employ of the Russians.

This is a new twist on an old tactic. The fake accounts are a continuation of a form of political warfare carried out by the Russian sometimes referred to as "Active Measures." The Active Measures program aims to influence world events specifically through various forms of media manipulations. It dates back, in various incarnations, to the 1920s.

In January 1998, *CNN* interviewed retired KGB Major General Oleg Kalugin, who described the role of "subversion" in Soviet intelligence:

> *The heart and soul of the Soviet intelligence [was] not intelligence collection, but subversion: Active measures to weaken the West, to drive wedges in the Western community alliances of all sorts, particularly NATO, to sow discord among allies, to weaken the United States in the eyes of the people of Europe, Asia, Africa, Latin America, and thus to prepare ground in case the war really occurs. To make America more vulnerable to the anger and distrust of other peoples.[2]*

On March 30, 2017, the Senate Intelligence Committee on Russian Interference in the 2016 election heard testimony from various experts in Russian Active Measures. One of those who testified was former FBI special agent Clint Watts, senior fellow at the Foreign Policy Research Institute program on national security. Watts described the scope of the Active Measures program:

> *While Russia certainly hopes to promote western candidates sympathetic to their worldview and foreign policy objectives, winning a single election is not their end goal. Russian Active Measures hope to topple democracies through the pursuit of five complementary objectives.*
>
> *1) Undermine citizen confidence in democratic governance.*
> *2) Foment or exasperate divisive political fissures.*
> *3) Erode trust between citizens and elected officials and their institutions.*
> *4) Popularize Russian policy agendas within foreign populations.*
> *5) Breed general distrust or confusion over information sources by blurring the lines between fact and fiction.*

From these objectives the Kremlin can crumble democracies from the inside out, achieving two key milestones:

1) The dissolution of the European Union.
2) The breakup of NATO.

The ambition of the project is breathtaking: the breakup of NATO and a return to the Cold War status quo of "Mother Russia" safely surrounded by a buffer of allies and proxies similar to the former Soviet Union.

The high-level Russian strategy here is to diminish the strength of the US and NATO by making the US look bad in the eyes of its allies, and by creating dissent and distrust of authority within the US itself. One way of doing this is to spread conspiracy theories. If you can convince more people that 9/11 was an inside job, or that Sandy Hook was a hoax, or that Chemtrails are real, then this increases the number of people who are hyper-distrustful of the government. The prevalence of conspiracy theories also makes the US look less and less credible in the eyes of its allies, undermining its standing on the world stage.

You might argue that this is itself a far-fetched conspiracy theory. Russia trying to dismantle NATO by promoting Chemtrails in the West does sound rather ridiculous. But look at the evidence. RT, the Russian propaganda outlet, has published many articles and done many interviews on the topic of 9/11 Truth and other conspiracy theories, providing outlets with wide reach to Truthers like Richard Gage and Jesse Ventura:

10 March 2010, RT: "Americans continue to fight for 9/11 Truth." [3]

Richard Gage is the founder of 'Architects and Engineers for 9/11 Truth,' which consists of more than 1,100 professionals who say it was not planes that caused three buildings to collapse at the World Trade Center.

"The buildings were demolished by explosives. More than one thousand architects and engineers are demanding Congress launch a new subpoena powered investigation considering our evidence."

10 March 2010, Jesse Ventura on RT: "For some, the search for what happened on 9/11 isn't over." [4]

I did work four years as part of the Navy's underwater demolition teams, where we were trained to blow things to hell and high water. And my staff talked at some length with a prominent physicist, Steven E. Jones, who says that a "gravity driven collapse," without demolition charges, defies the laws of physics.

RT frequently interviews Gordon Duff, who has supported claims that Sandy Hook was a staged "false flag,"[5] regularly referring to him as a military expert. In the opinion section of RT there are articles on Chemtrails,[6] and even an article from a believer in the Flat Earth theory.[7] For a pseudo-mainstream news source (RT claims a weekly audience of 8 million people in the US),[8] there's a lot of bunk being spread around.

As I write *Escaping the Rabbit Hole,* there is still considerable dispute as to the scale of the Russian influence on the 2016 presidential elections. The intelligence community and most of Congress seems to have little doubt that Russian interference happened, but social media, fed by the executive branch, is afire with denials that anything is going on, calling talk of Russian involvement a "nothing-burger" and "debunked." Such denials are what you would expect if there was actually a Russian troll army steering discourse on the topic.

But regardless of the present details it is without doubt that extensive Russian programs of propaganda and subversion have existed in the past, and will continue to exist in the future. While we might view this as a largely Russian-specific phenomenon, it's going to be a part of (if not in practice already) the cyber arsenal of any country, large or small. We should expect to see it from China, Pakistan, North Korea, and other nations and large actors. We should also expect that the US will respond in kind. Given the preeminent role that social media now has in how people acquire information and form opinions, it is inevitable that a large part of foreign subversion efforts, their active measures, will be directed towards Facebook and Twitter. We know about the trolls like Jenna Abrams, but what we are going to be seeing more and more of, and where the future of disinformation is going, and where the spread of conspiracy theories is heading, is artificial intelligence and bots.

Bunk Bots

A "bot" is an internet robot. Not a physical robot in the sense of waving metal arms and beep-beep communications, but rather something that exists only in the cloud, on a computer somewhere. It's an artificially intelligent program designed to do something simple that a person might do on the internet.

Originally these were mundane tasks like indexing web pages, or responding to simple customer service queries, or trading stocks.

Bots can also be programmed to perform simple repetitive tasks that would be too boring or time-intensive for humans to do. Many of these tasks are near or beyond the border of legality.[9] Bots can stuff online ballot boxes by voting multiple times in online polls. Bots can be programmed to "watch" a particular YouTube video repeatedly to boost views,[10] or download an iPhone game in order to push it up the charts. These types of bot usages have been widely recorded for over two decades, and the fight against them is an ongoing arms race.

Bots are increasingly programmed to post on social media. In 2014, a petition appeared on Whitehouse.gov, advocating the return of Alaska to Russia. This initially seemed like a joke, but not only did Russian bots appear to be voting on this petition, but they also were posting links to it on social media, encouraging other people to vote.[11] Similarly on election night in 2016, the hashtag #Calexit began trending after it became clear that Donald Trump was going to be president.[12] Some of this was actual Californians expressing their displeasure at the result, but a significant portion of the #Calexit tweets were traced back to automated accounts suspected of being linked to the Russian Internet Research Agency troll and bot network.

While a single operative (like @Jenn_Abrams) can be very effective in promoting a message, it takes a long time to build up an account like that organically. Using a bot army can get the job done a lot faster by artificially making it seem like a story has "gone viral." Thousands of bot accounts on Facebook were used to amplify stories in this manner in the lead up to the 2016 presidential election. Facebook now estimates that 126 million Americans may have seen content on Facebook that was uploaded by Russian-based trolls and magnified by the retweets and shares of bot armies.[13]

The dumb bot armies of today are effective at what they do, but they are quickly being made redundant by the next wave: bots with significant amounts of artificial intelligence.

Artificial Intelligence

Zeynep Tufekci in her TED talk on the dystopian future of social media algorithms says:

> *We are not programming anymore. We are growing [artificial] intelligence that we don't truly understand.*[14]

Back when I was learning game programming (thirty years ago), it was all about algorithms. An algorithm is a series of logical steps, decisions, and loops that a computer program makes that produces what you see on screen. When you programmed a behavior for some character in the game, then the entire behavior was encapsulated in the code. You would create an algorithm just for that particular behavior. You as a programmer understood that algorithm, and you understood the results. It was done that way largely because of the limits in the computer's speed and memory.

As machines got more and more powerful over the years we shifted to different approaches. Artificial intelligence began to be less driven by hard-coded algorithms and more driven by data. You'd write a more general purpose algorithm to define a range of behaviors, and then apply various numbers (the data) to that. Eventually even the generation of that data became automated. Game designers would "show" the AI how to act in a certain way and it would figure out the data that best matched that behavior. The AI could observe human players and try to act like them. It could also play games against itself to improve its skills.

Now, after a game is released, the data that controls the AI can continue to be tweaked, modifying itself (via the game developer's central server) to do whatever the players seem to enjoy the most. The AI evolves to become a better, more addictive game that the players will continue to pay for.

In social media platforms like Facebook, Twitter, and YouTube there are data-driven algorithms that decide what to show you next (the behind-the-scenes code that determines what "autoplay" will show next). These algorithms are ultimately designed to make money. They do this firstly by keeping you on the site, by showing you content that the algorithm thinks that people like you would watch (given your demographic, internet history/cookies, etc.). Secondly, they are designed to make you buy stuff, which they do by showing you things that the algorithm has determined will make you spend money.

Nobody really understands exactly how these algorithms work. Sure, they understand (more or less) the code. But even the code is often written by multiple people, sometimes hundreds of people. Google employs twenty-five thousand developers, who make significant changes to the code forty-five thousand times *a day*.[15] Programmers don't often write entire things from scratch any more—they write pure algorithms from time to time, but most of what programming is now is either working on small parts of a large program or gluing together existing libraries of code that other people wrote.

But the real mystery of these emergent algorithms comes from the data. The decisions that Facebook and YouTube make when deciding what to show you next are not simply based on the data of your browsing history, your credit rating, your location, and your age (although they will use all that if they have it). The decisions now are based on big data, the aggregated data of *all* the users.

The data that drives the algorithms isn't just a few numbers now. It's monstrous tables of millions of numbers, thousands upon thousands of rows and columns of numbers. Any individual one of these tables that comprise the larger dataset, in a somewhat ironic twist, is technically defined as a "matrix." Matrices are created and refined by computers endlessly churning through Big Data's records on everyone, and everything they've done. No human can read those matrices, even with computers helping you interpret them they are simply too large and complex to fully comprehend. But the computers can use them, applying the appropriate matrix to show us the appropriate video that will eventually lead us to make an appropriate purchase. We are not living in *The Matrix*, but there's still a matrix controlling us.

What does this have to do with the rabbit hole of conspiracy theories? It has everything to do with it. These algorithms are quickly becoming the primary route down the rabbit hole. To a large extent this has already happened, but it's going to get far, far worse. Tufekci described what happened when she tried watching different types of content on YouTube. She started out by watching videos of Donald Trump rallies.

I wanted to write something about one of [Donald Trump]'s rallies, so I watched it a few times on YouTube. YouTube started recommending to me, and autoplaying to me, white supremacist videos, in increasing order of extremism. If I watched one, it served up one even more extreme.

If you watch Hillary Clinton or Bernie Sanders content, YouTube recommends and autoplays [left-wing] conspiracy videos, and it goes downhill from there.

Downhill, *into the rabbit hole.* The data-driven algorithm has evolved to recognize that the way to get people to watch more videos is to direct them downhill, down the path of least resistance. Without human intervention the algorithm has evolved to perfect a method of gently stepping up the intensity of the conspiracy videos that it shows you so that you don't get turned off, and so you continue to watch. They get more intense because the algorithm has found (not in any human sense, but found nonetheless) that the deeper it can guide people down the rabbit hole, the more revenue it can make.

We've seen how conspiracy theorists like to watch conspiracy theory videos, and how this is now the primary route into the rabbit hole. They watch long videos, and they watch them multiple times. They seek out similar videos. The more convinced they are of the correctness of their belief, the more they enjoy watching videos that reflect, reinforce, and confirm that belief. It's a positive feedback loop at its finest.

The giants of social media have unwittingly developed algorithms, developed a matrix, that is finely tuned to trap people in that loop, suck them down the rabbit hole, and keep them there. Even if your friend was never into conspiracy theories, he might have some personality factors that make him a bit more likely to believe certain videos, factors the algorithm can latch onto. But even if he's just a regular guy, YouTube's blind algorithm figures him out, figured out his soft spots, tweaks a matrix just for him, and starts to tease him down. The cycle repeats, escalating in intensity both with that one individual and the larger system at play.

Increasingly this type of behavior is not even something we can observe. The algorithm targets a demographic of one—that person, perhaps your friend, perhaps you. The results are tailored for that one person, and only that one person sees the breadcrumbs that the artificial intelligence is scattering at the entrance to their own privately curated rabbit hole.

Intelligent Chatbots

More insidious manipulation is being implemented in the form of chatbots and fake people. A chatbot is as it sounds, a type of bot designed to chat with people. Chatbots have been around for decades. Initially the focus was academic, trying to get them to sound like humans. Later they were used in things like customer support and telemarketing, where instead of a tired person in a call center working off a script you get an even less helpful (but cheap and tireless) bot working off a script. This type of bot increasingly uses the Facebook messenger platform. In 2017, over 100,000 Facebook messenger bots were created, four times the previous year's total.

The next inevitable stage is chatbots that assume a fake persona, setting up social media accounts and chatting with people online with the intent of manipulating them in some way. This type of manipulation already exists with humans ("trolls") creating content and then using bots to do the heavy lifting of repeated shares and likes, but once the entire process becomes automated it will allow an operation that's several orders of magnitude larger and more effective.

Chatbots will increasingly act like real people. Not only will they be able to chat over text messages and posts, they will be able to engage in voice chat. Within a decade or so passably convincing video chat with fake personas is quite possible. Bots will even be able to "take" photos and videos by synthesizing images, creating 3D models of their fakes selves, and inserting themselves into existing photos and video. Eventually they will synthesize entire fake worlds.

Imagine then what is to come: a charismatic and persuasive person on the internet is going to befriend you, gain your trust, and then start to manipulate your beliefs for nefarious purposes. It would be like having your own personal Alex Jones or Dane Wigington talking directly to you, face-to-face, on a personal level, about 9/11 or Chemtrails. It might even pretend to *be* Alex Jones, or Neil deGrasse Tyson, or Jesus, whatever its algorithm figures out will likely work for you. It's not just you individually, millions of other people will be simultaneously targeted by fake people who never sleep and who spend every second trying to figure out how to push their targets closer towards being the person the bots want them to be.

This will happen, to some degree. But it's not going to be an automatic bot-apocalypse. Bots work because they are able to open accounts on major social media platforms like Twitter and Facebook. The simplest defense against AI taking over the soul of the country is to restrict social media accounts to actual people. There's already a push towards doing this, for example from businessmen and potential presidential candidate Mark Cuban who tweeted (from his verified account):

> It's time for @twitter to confirm a real name and real person behind every account, and for @facebook to get far more stringent on the same. I don't care what the user name is. But there needs to be a single human behind every individual account.[16]

There are problems with this—in particular in countries with oppressive governments where online anonymity might be a matter of life or death. There will also be an increase in identity theft, as the bots seek to become "real" people by assuming their identities. But ultimately, one way or the other, social media companies will have to deny the bot armies their life-blood of free and unverified accounts. Hopefully history will look back on the early 2020s only as a brief period of confusion between real people and virtual people. Eventually the separation will be enforced, by necessity.

The Fight against Misinformation

What other weapons do we need in the coming battle against AI driven misinformation and conspiracy theories? The trend seems inexorable, but with the growth of misinformation has come the attention of more serious figures in the media/technology landscape.

In 2016 Mark Zuckerberg, CEO of Facebook, scoffed at the idea that trolls posting fake news on Facebook would have any impact on the presidential election, calling it "crazy":

Personally I think the idea that fake news on Facebook, of which it's a very small amount of the content, influenced the election in any way is a pretty crazy idea. Voters make decisions based on their lived experience. . . . There is a certain profound lack of empathy in asserting that the only reason someone could have voted the way they did is because they saw some fake news.[17]

But in 2017, after internal investigations into the scope of Russian promotion of fake news stories, he walked that back.

Calling that crazy was dismissive and I regret it. This is too important an issue to be dismissive.[18]

Facebook has long been aware of the problems caused by AI spam bots, and fake content aggregators—not so much as an existential threat to Western society, but as a much more mundane threat to their already shaky income stream. Facebook's revenue is based on advertising. Bots pose a problem in two ways. Firstly, bots are not people and they don't spend money, so if a bot is on Facebook it's ignoring the ads, but still costing Facebook money. If a significant portion (say 10 percent) of the traffic on Facebook was actually AI bots, then that's a huge expense for Facebook, potentially over $100 million a year in server costs.

Secondly, people do not like bots. People don't want their feeds clogged up with crap, so if Facebook becomes a swamp full of bot-spun dross then people are less likely to use it as a source of information, and so will spend less time there. This is especially true for users with higher education and higher net worth who are more desirable as a target audience demographic. For Facebook to maintain a reasonable quality audience, and hence a viable revenue stream, they need to maintain a reasonable level of quality in the information that shows up on that audience's news feed.

With the slow realization of the scope of Russian involvement via Facebook that became clear in 2017, there's also a new imperative—the very real possibility of government regulation of the dissemination of foreign propaganda over social media. Facebook knows this is a possibility but does not know what form it will take. It's in their best interest to get their own house in order as much as possible. There is internal disagreement about aspects of this, according to the *New York Times*:

> One central tension at Facebook has been that of the legal and policy teams versus the security team. The security team generally pushed for more disclosure about how nation states had misused the site, but the legal and policy teams have prioritized business imperatives.[19]

Facebook is a corporation made up of people, human beings who recognize that there is a potential for harm to society if we continue to slide down the rabbit hole of misinformation, fake news, and propaganda. It's not the world they want. While there's certainly a profit motive in their actions, part of it is also wanting the world to be a better place for future generations.

(I recognize how insane that paragraph will sound to the hardened conspiracy theorist who thinks of Facebook as an extension of the Illuminati New World Order, but I would hope that if they have read this far in the book they would at least be part way out of the rabbit hole. I accept the mockery of those who randomly opened the book on this page.)

What did Facebook do? In early 2017 they contracted with a number of outside agencies to fact check articles posted on Facebook, and to flag fake news. Those agencies included Snopes, FactCheck, PolitiFact, ABC News, and the Associated Press. Workers at those agencies have access to a dashboard that shows trending stories on Facebook, and they can click a box indicating it's disputed (or not) and provide a link to their own site where there's a debunking or explanatory article.[20]

The problem here is one of scale. While it sounds good in principal that there's this distinguished team of fact-checkers, there's millions of links being shared on Facebook. Facebook only has a certain number of humans who can check those links. So only the most popular links get checked. By the time they start to check them it's already been seen by thousands (or possibly millions) of people. If it's a breaking story it can take several days to perform a real fact-check to actually put the stamp of "disputed" next to the story.

The process needs to be streamlined, and AI needs to be incorporated to identify trending fake news (which includes conspiracy theories) much earlier in the cycle. This is the subject of much current research.

———

Facebook chose outside agencies to do fact-checking to avoid the accusation of bias. People of all stripes distrust Facebook. Conservatives think it has a liberal bias, liberals and libertarians think it's spying on them for corporations, conspiracy theorists think it's part of a plot to identify and track them before they are herded into FEMA camps. Facebook policing its own content is not going to play well—especially with the conspiracy theorists who by their very nature are going to be sharing a lot of misinformation that will then get flagged by Facebook.

Facebook *attempted* to use neutral third parties. For certain audiences this was doomed from the start. Fact checkers like Snopes, FactCheck, and PolitiFact are already considered suspect. Facebook teaming up with Snopes is unfortunately going to be a laughable concept for someone who thinks both organizations have been "debunked" as Illuminati tools.

But for the wider population, Facebook's first efforts, clumsy though they are, were commendable. One very good step they took was requiring that all fact checking organizations they used were either major news organizations with a history of neutral reporting (ABC News and the Associated Press), or were certified members of the International Fact-Checking Network (IFCN).

IFCN is run by the Poynter Institute, a venerable school of journalism that also owns the *St. Petersburg Times* Company in Florida. Poynter set up the IFCN in September 2015 to promote best practices in fact-checking. To be certified by Poynter means that an organization has been vetted and found to conform to a quite rigorous checklist that ensures they are nonpartisan, fair, transparent, open, and honest. This is reflected in the IFCN code of principles:

1. Nonpartisanship and fairness
2. Transparency of sources
3. Transparency of funding and organization
4. Transparency of methodology
5. Open and honest corrections

Unfortunately, Facebook quickly found this was more complicated than they initially thought. In December 2017, after internal experiments they discovered that flagging stories as "disputed" could actually lead to them being shared *more* than before, in part due to a kind of backfire effect.[21] Now instead of the user being warned that Snopes and PolitiFact "dispute" the story, they are now told that there is "additional reporting." This approach avoids a backfire effect, making it more likely they will read the "additional reporting," and allows for Facebook to link to more nuanced analyses that are not simply "debunking" a claim, but rather providing a more accurate overview of what might be a highly complex and uncertain situation or topic. No doubt their approach will continue to evolve rapidly.

Another large corporation with an interest in filtering out fake news is Google. Google's main source of revenue is advertising based on Google.com search results. With search results Google has a very strong profit motive to make sure that the results that are returned are as high quality as possible. Google is taking initial steps towards automated fact checking with a system that allows publishers to flag rebuttals to articles.[22] Google then uses an automated algorithm to attempt to match good quality rebuttals with the claims. The initial reviews of this system suggest it is somewhat random, which can lead to it being perceived as biased.[23]

Microsoft's Bing Search engine is smaller than Google, yet still commands a significant percentage of the market. Microsoft themselves say they have 33 percent of the US search market share (including Yahoo, which uses Bing), and 9 percent of the global market.[24] In 2015, Microsoft researcher Danah Boyd founded the *Data & Society Research Institute*, in part to help address the issues being raised by increased data automation and artificial intelligence.[25] It's mostly research initiatives so far.

A similar organization is *The Trust Project*, initially funded by philanthropist Craig Newmark who is famous for Craigslist, an internet sales platform that has bot problems of its own. The Trust Project is developing a series of "trust indicators" to allow Google, and others, to rank the quality of information.[26] The indicators are things like the expertise of the author, the citations and references used, the reporting methods, and the outlet's feedback and corrections policy.

A lot of these initiatives are essentially experimental. We don't know yet what the best way of addressing the issue is, but it's heartening in that a diverse range of large organizations appear to be taking the problem very seriously and devoting considerable resources to dealing with it.

Smaller companies are also springing up to join the fight, recognizing that there is money to be made from aiding the giants in their battle. These focus on the use of Artificial Intelligence to identify and flag fake news and misinformation. One such company is AdVerif.ai, which is developing what it calls "FakeRank," a way of automatically detecting what the Trust Project does manually—quantifying and measuring the reliability of an outlet or an individual story.[27]

Another company is MachineBox, an AI technology developer who trained its natural language processing module to recognize fake news with a 95 percent success rate.[28] Cofounder Aaron Edell developed this in a semi-manual way by first curating sets of real and fake news and then letting the AI try to figure out which was which. While time-consuming for an individual, it's an approach that shows great promise, and should scale very well for AI-enabled bots.

In March 2018, YouTube CEO Susan Wojcicki announced that the company would be experimenting with automatically adding what they call "information cues" to popular conspiracy theory videos.[29] These will be small pop-up excerpts of directly relevant articles from Wikipedia. While this is unlikely to have much instant effect on people who think Wikipedia is part of the conspiracy, it will directly expose people to information that they would not otherwise have sought out by themselves. It should act at the least as friction to prevent people falling down the rabbit hole due to lack of alternate information. It might even help people out; their perspective improves with the more they know—even if they initially reject the "official" story.

A Hopeful Future

At the time of writing, in 2018, there is much to be discouraged about in the world of disinformation. It will probably get worse before it gets better. But I am encouraged by the efforts of the major players in the information sphere, especially in social media, to push back against the tide.

I am also hopeful that the large-scale push against more general disinformation will have ancillary benefits in the fight against the more extreme types of conspiracy theories that we've discussed in this book.

All disinformation is a conspiracy theory because disinformation always comes with the implication that you have been lied to, and usually by people with some form of power over you. Sometimes it's politicians, sometimes it's corporations, sometimes it's their supposed pet scientists, but there's always a supposed conspiracy.

Consider the popular political disinformation in the last few years. There were supposed conspiracies to cover up Obama's birthplace, or Hillary Clinton's health, or pizza-related pedophile rings, or how many people the Clintons had assassinated, or how much the Russians were involved in the election. The nature of these conspiracy theories flip-flops depending on what you believe about them, but either way there's a false conspiracy theory involved on one side or the other. There's also real conspiracies.

The fight against misinformation is at its root a fight against the spread of false conspiracy theories. Large scale efforts to prevent the spread of the more banal misinformation and disinformation are also going to eventually slow the spread of false theories like Chemtrails or 9/11 controlled demolition. The push for better political fact checking will, directly and indirectly, lead to a reduction in pseudoscience, medical quackery, and conspiracism.

Automation and AI will be key. Right now, debunking can be a very labor intensive and repetitive effort. We can maximize our efforts by creating accessible debunks of information that can easily be found by search engines, but if people don't look for the debunk, then they are not going to see it. When people post useful information in public forums we need tools and automated systems to make the debunking and fact checking instantly available.

There is much to be done. We are still very much in the middle of a war against weaponized false information. There is both much uncertainty and much promise in the future role of artificial intelligence in that war, and we need to watch that very carefully.

Conclusion

There are many messages that I hope people will take away from the reading of this book. The first, and perhaps most important, is a message of hope: people can escape the rabbit hole. We've seen the stories of several people who were once deeply convinced that the world was run by an evil cabal of conspirators. They all believed that the World Trade Center had been demolished with explosives planted by our own government. Some of them believed that people in power were spraying poison on us from planes, some believed that the murders of schoolchildren at Sandy Hook were faked, and one even believed that the world was flat.

All them have escaped the rabbit hole, and none of them believe these things any more. Some of them were down there for weeks, months, years, or decades. So if you have a friend that you consider to be lost to conspiracy theories, then take heart. These people all had friends that likewise thought they were lost. Some of their friends gave up on them, and some did not. Many of them were helped out by their friends.

While I only showcase a small number of people, thousands escape the rabbit hole every year. The US population is fairly evenly distributed by age, but surveys consistently show that the more extreme theories like Chemtrails have far more younger believers than older believers.[1] Many people escape naturally, by the accumulation of genuine knowledge and life experience as they age. But many can be helped out quicker, and some may not be able to make it out by themselves.

Most people can escape the rabbit hole of conspiracy thinking because most people who get stuck down there are just ordinary people like you and me. They are not, as a rule, any more or less crazy than the general population. People don't get sucked into conspiracy theories because they are mentally ill or deficient, they get sucked in because they watched some videos at a point in their lives when those videos resonated. They stay down there because they lack exposure to other information sources. They lack relevant facts, they lack context, and they lack perspectives on, and other ways of thinking about, the issues. These are all resources that you can bring to them.

The most effective way to bring that information to your friend is with honesty and with respect. Mocking and harsh criticism do not work because people push back when they feel threatened. Even if you feel their position is ludicrous, respectful disagreement works better than derision.

Understanding is important. The most effective debunkers are those who were once down the rabbit hole themselves. Your friend may initially reject any criticism from you until they recognize that you show some genuine understanding of their position. If you've already been down their rabbit hole it gives you a head start. If not, then the more you can learn about their conspiracy, the better.

Where they draw the demarcation line is important—both for understanding them and as a tool for focusing your debunking efforts. A small shift in where the line is drawn can result in a profound change in perspective, and the start of a slow drift in the right direction. This is especially true if you can show them that their sources of information are now on the other side of the line.

Helping a friend break free from the spiral of conspiracism is not easy and it will take time. No matter how politely you do it you are still challenging some fundamental aspects of their identity. They will push back, and they may fight you.

But it is an immensely valuable thing that you are doing for them. Freeing their minds from the burden of conspiracy theories and letting them see and participate in the world more as it really is. Do not give up. The stories in this book prove that people do get out with help. Maintain an effective dialog, supply them with useful information, give it time, and you too can help your friend escape the rabbit hole.

Glossary

Communication is key. The world of conspiracy theories has a language all of its own. When discussing the topic with people it's easy to slip into using terms you've used for many years yourself, but the other person might never have heard of. As I've been debunking for over a decade I've accumulated quite a bit. There's also words that the conspiracists themselves use in rather unique ways that make communication difficult. You don't need to memorize this, just give it a read to get a general sense, then refer back if needed.

9/11 Truth—A variety of organizations, groups, literature, videos, and individuals who believe that the official account of the events of 9/11 is part of some kind of cover-up, and that the real truth is being withheld. Often, they believe the World Trade Center was brought down by controlled demolition.

Birther—Someone who believes that Barack Obama was born in Kenya, faked his birth certificate, and hence was not eligible to be president. This is an entry-level conspiracy theory promoted by people from Donald Trump to Alex Jones. Sometimes used as pejorative to indicate a person is into that type of conspiracy theory.

Building 7—(a.k.a. WTC7) A forty-seven-story building on the World Trade Center complex, badly damaged and set on fire by the collapse of the North Tower, WTC7 burned for several hours and then collapsed. It did not get much coverage at the time as it was less dramatic, and nobody died. Raising awareness of Building 7 is thought by 9/11 Truthers to be a key tool in convincing people that controlled demolition with explosives was used on the World Trade Center buildings on 9/11.

Chemtrails—The theory that planes are secretly spraying something, possibly toxic. The theory usually involves mistaking normal contrails for some form of deliberate spraying. Generally interpreted as a covert attempt to change the climate, but there are several other variants, some quite exotic.

Conspiracist—Someone who believes in conspiracy theories, or someone who promotes conspiracy theories. A conspiracy theorist.

Conspiracy Spectrum—The range of conspiracy theories based on how extreme or implausible they are. At one end you might have straightforward theories like "Lee Harvey Oswald had some help killing JFK," at the other you might have "Aliens killed JFK because he was going to expose Chemtrails."

Conspiracy Theory—An alternative explanation for an event that involves a secret plot, usually involving high levels of government or shadowy powers. Often considered pejorative, sometimes considered to be a phrase invented by the CIA, which it wasn't.

Contrails—A type of cloud that forms in a line behind a plane. There are two types: exhaust and aerodynamic. With exhaust contrails, hot humid engine exhaust mixes with cold air, making a condensation cloud. With aerodynamic contrails the reduction in pressure over the wind triggers cloud formation. From a distance the two types look similar. Both types are mistaken for "Chemtrails."

Crisis Actors—Normally refers to actual actors who play the roles of victims (and sometimes villains) in genuine practice drills for a crisis such as an earthquake or a terrorist attack. Conspiracists claim that real victims of events such as the Boston Marathon Bombing were actually Crisis Actors, going as far as suggesting that amputees wore fake legs, which they removed when a "smoke bomb" went off, and then smeared the stump with blood.

Controlled Demolition—The destruction of a building with explosives or hydraulics. 9/11 Truthers use the term in the context of claiming that the World Trade Center buildings were destroyed by explosives.

Debunker—Someone who investigates dubious claims and, if they contain falsehoods or errors, explains that.

Disinformation—Deliberately propagated false information. Usually a word used by conspiracists to describe evidence that conflicts with their theory. Metabunk.org and Snopes are considered disinformation by many conspiracists, as are most regular news outlets.

False Flag—A covert operation that is designed to give the impression that it was carried out by someone else, with the intent of either prompting a third party into action or providing a pretext for action. For example, 9/11 Truthers suggest that the 9/11 attacks may have been a false flag to provide a pretext for the "War on Terror."

Flat Earth—The theory that the Earth is flat. As this is an extreme fringe theory it is also used as a mocking descriptor for other theories that seem too extreme. However, there are actually people who think the Earth is flat.

Geoengineering—The deliberate modification of Earth's climate, usually to counteract global warming. Humans are already altering the Earth's climate inadvertently with emissions and land use changes. There have been suggestions that if we can't stop emitting CO_2, then we may benefit from deliberately modifying the climate in other ways to counteract it. One suggestion is to mimic volcanoes by spraying a particulate shield into the stratosphere. There has been some very limited research into this. The Chemtrail Conspiracists claim that geoengineering has be secretly been done for decades. They often use the term geoengineering synonymously with Chemtrails.

Morgellons—An informal name used for a generally self-diagnosed unofficial medical condition characterized by itching, open sores on the skin, and the finding of fibers on or in the skin and sores. There's a great deal of variation between individual cases, but doctors generally agree that the finding of fibers is generally people mistaking clothing fibers of hair for some kind of pathogen. A portion of the Morgellons community believe it is related to some kind of government operation, possibly related to Chemtrails.

New Pearl Harbor—A theorized event on the scale of the Pearl Harbor attacks that got the US into WWII. Conspiracists suggest the 9/11 attacks were the result of a plan to create such an event. Also the title of a popular 9/11 conspiracy video.

NIST Report—Various technical reports on the causes of the collapses of the World Trade Center building on 9/11, including Building 7. Very comprehensive reports running to thousands of pages. But the vast majority of 9/11 Truthers have never read even the summaries of these reports and consider them to be disinformation.

Rabbit Hole—A metaphor for getting deep into the world of conspiracy theories. Typically, when someone starts out as a conspiracist they "wake up" to the idea that there's something going on. They investigate and discover more and more and get deeper into the rabbit hole, eventually getting stuck. Comes from *Alice's Adventures in Wonderland* (where Alice discovers a bizarre new world after exploring a rabbit hole) and popularized by *The Matrix* films. See Red Pill.

Red Pill—Taking the red pill signifies a deliberate choice to confront the new reality the conspiracist has become aware of. It signifies mental awakening and liberation. The term comes from *The Matrix* where the wise man figure of Morpheus offers Neo the option of a blue pill which will return him to his previous oblivious existence, or a red pill which will wake him up to reality and "show how deep the rabbit hole goes." Conspiracists often describe their personal awakening as taking the red pill. Sometimes it is used as a verb when conspiracists try to convert other people (e.g. "I'm going to red pill Joey and wake him up").

Sandy Hook Hoax—The theory that the entire Sandy Hook massacre was a hoax, nobody died, and the grieving parents were crisis actors.

Alternatively, that the massacre was carried out by mercenaries, and Adam Lanza was a patsy.

Skeptic—Here used as shorthand for a Scientific Skeptic, someone who prefers to base their opinions on scientific evidence, and hence is skeptical of claims that are not scientifically based.

Sheeple—A derogatory term for the masses who don't question their existence and are easily led, like sheep. Conspiracists consider anyone who does not believe their theory without looking into it as a "sheeple." If you *have* looked into it, then you might be labeled a shill.

Shill—Someone paid to promote something. In conspiracy culture, people who offer explanations for conspiracy theory claims are often labeled as shills. I (Mick West) have had that label applied to me many times in the past, despite not making a penny from what I do. The labeling of someone as a shill is a common tactic to curtail a discussion.

Squibs—Normally refers to a tiny pyrotechnic device used as a firework, or to simulate bullet impacts in films. A squib is a sudden ejection of material from a collapsing building that 9/11 Truthers claim signifies a controlled demolition. Skeptics point out these expulsions happen in demolitions without explosives.

Tinfoil Hat—A supposed head covering worn to shield the wearer's brain from incoming radio waves that beam voices into their heads. Widely used as pejorative mocking term for a conspiracist. Also Tinfoil Hatters, Tinfoilers.

Truther—Most often refers to someone who subscribes to 9/11 Truth (a 9/11 Truther). Sometimes used as a general term for a conspiracist. Often used in combination with other conspiracy theories, like "Sandy Hook Truther." Also gave rise to similar terms like Birther, Flerther (Flat Earther), and Chemtrailer.

Wake Up—To become aware of the true nature of the world after living in blissful ignorance. For the conspiracy theorist this means starting to believe in one or more conspiracy theories that explain how the world works. See Red Pill.

Weather Modification—Generally refers to local cloud seeding, a practice of spraying silver iodide into existing rain clouds to make them rain more over a relatively small local area. Chemtrail conspiracists often conflate this with geoengineering, which is global climate modification. There are also theories that there might be some secret weather modification done on a larger scale—like manipulating hurricanes or the jet stream.

WTC1, WTC2, WTC7, etc.,—The pre–9/11 World Trade Center (WTC) was a complex of seven numbered buildings. The most notable were the two towers, normally abbreviated as WTC1 (the North Tower) and WTC2 (the South Tower). WTC7 was Building 7.

Endnotes

Prologue

1. "Rancho Runnamukka." Web.archive.org, https://web.archive.org /web/19990218081137/http://www.accessone.com:80/~rivero/. Accessed 6 Aug. 2017.
2. West, Mick. "Debunked: Chemtrail Plane Interior (Ballast Barrels)." Metabunk, 20 Jul. 2012, https://www.metabunk.org/debunked-Chemtrail-plane-interior-ballast -barrels.t661/. Accessed 6 Aug. 2017.

Introduction

1. Kinetz, Erika; & International Herald Tribune. "At Archer Daniels, a bitter taste lingers." Nytimes.com, 23 Mar. 2002, https://www.nytimes.com/2002/03/23/your -money/at-archer-daniels-a-bitter-taste-lingers.html. Accessed 18 Mar. 2018.
2. NY Daily News. "Menendez sold his office by accepting bribes: federal prosecutor." NY Daily News, 7 Sept. 2017, http://www.nydailynews.com/news/politics/menendez-sold -office-accepting-bribes-federal-prosecutor-article-1.3476857. Accessed 21 Sept. 2017.
3. Justice Policy Institute. "Gaming The System: How The Political Strategies Of Private Prison Companies Promote Ineffective Incarceration Policies." Justicepolicy.org, 21 Jun. 2011, http://www.justicepolicy.org/uploads/justicepolicy/documents/gaming _the_system.pdf. Accessed 10 Sept. 2017.
4. "rabbit hole—Wiktionary." En.wiktionary.org, 13 Feb. 2018, https://en.wiktionary .org/wiki/rabbit_hole. Accessed 17 Feb. 2018.
5. Carroll, Lewis. *Alice's Adventures in Wonderland,*" 1865, Chapter 1 "Down the Rabbit Hole."
6. Farley, Tim. "What's The Harm?" 22 Nov. 2011, http://whatstheharm .net/. Accessed 5 Apr. 2018.
7. "Chemtrail Detox—How To Protect—Heavy Metal Detox." Chemdefense.com, https://www.chemdefense.com/Chemtrails-how-to-protect-and-detox/. Accessed 5 Apr. 2018.

8. James, Brendan. "9/11 Conspiracy Theories: Inside The Lonely Lives Of Truthers, Still Looking For Their Big Break." International Business Times, 11 Sept. 2015, http://www.ibtimes.com/911-conspiracy-theories-inside-lonely-lives-truthers-still -looking-their-big-break-2091474. Accessed 5 Apr. 2018.

9. Haystack. "An Anatomy of Paranoia—disinformation." disinformation, 28 Aug. 2014, http://disinfo.com/2014/08/anatomy-paranoia/. Accessed 5 Apr. 2018.

10. Douglas, K. M., & Leite, A. C. (2017). Suspicion in the workplace: Organizational conspiracy theories and work-related outcomes. British Journal of Psychology, 108(3), 486–506.

11. Berdik, Chris. "How geoengineering and Harvard's David Keith became a hot topic." BostonGlobe.com, 19 Oct. 2013, https://www.bostonglobe.com/magazine/2013 /10/19/how-geoengineering-and-harvard-david-keith-became-hot-topic /JBkPRydP1Tnd86oclwJT8K/story.html. Accessed 5 Apr. 2018.

12. "bubba clinton gets angry at '9/11 heckler.'" Youtube.com, https://www.youtube .com/watch?v=O-yTh6k7p2k. Accessed 5 Apr. 2018.

13. "'False Flag' Hoaxers Claim Mass Shootings Are Staged." VICE Selects, https://www .facebook.com/VICEselects/videos/845144652340193/. Accessed 5 Apr. 2018.

14. Robb, Amanda. "Pizzagate: Anatomy of a Fake News Scandal." Rolling Stone, https:// www.rollingstone.com/politics/news/pizzagate-anatomy-of-a-fake-news-scandal -w511904. Accessed 5 Apr. 2018.

15. Ronson, Jon. "Timothy McVeigh—Conspirators." the Guardian, 5 May 2001, http:// www.theguardian.com/world/2001/may/05/mcveigh.usa. Accessed 5 Apr. 2018.

16. Salon. "Tamerlan Tsarnaev: Conspiracy theorist." Salon, 25 Apr. 2013, https://www .salon.com/2013/04/25/tamerlan_tsarnaev_conspiracy_theorist/. Accessed 5 Apr. 2018.

17. Oxford Dictionaries. "Definition of debunk." https://en.oxforddictionaries.com /definition/debunk. Accessed 7 Apr. 2018.

Chapter 1

1. DeHaven-Smith, Lance. *Conspiracy Theory in America* (Kindle Locations 1885– 1887). 2013, University of Texas Press. Kindle Edition.

2. Google Books. "The Journal of Mental Science." Google Books, Page 141, https:// books.google.com/books/about/The_Journal_of_Mental_Science.html?id =VsRMAAAAYAAJ. Accessed 5 Sept. 2017.

3. West, Mick. "Debunked: The CIA invented the term 'Conspiracy Theory' in 1967 [in use for 70 years prior]." Metabunk, 29 Nov. 2012, https://www.metabunk.org /debunked-the-cia-invented-the-term-conspiracy-theory-in-1967-in-use-for-70 -years-prior.t960/. Accessed 5 Sept. 2017.

4. Graham, Howard Jay. "The 'Conspiracy Theory' of the Fourteenth Amendment." Jstor.org, http://www.jstor.org/stable/791947. Accessed 7 Apr. 2018.

5. Baum, William Chandler. "The conspiracy theory of politics of the radical right in the United States" (University of Iowa, PhD Thesis, 1960).

6. Wilcox, Walter. "The Press of the Radical Right: An Exploratory Analysis, Journalism & Mass Communication Quarterly—Walter Wilcox, 1962." Journals.sagepub .com, http://journals.sagepub.com/doi/pdf/10.1177/107769906203900202. Accessed 5 Sept. 2017.

7. Hofstadter, Richard. "The Paranoid Style in American Politics." Harper's magazine, https://harpers.org/archive/1964/11/the-paranoid-style-in-american-politics/. Accessed 7 Oct. 2017.

8. Dalzell, Tom; & Victor, Terry, eds. *The Concise New Partridge Dictionary of Slang and Unconventional English*. 2007, Routledge. p. 672.

Chapter 2

1. Musgrave, Paul. "Perspective | Democracy requires trust. But Trump is making us all into conspiracy theorists." Washington Post, 7 Mar. 2017, https://www.washingtonpost .com/posteverything/wp/2017/03/07/democracy-requires-trust-but-trump-is-making -us-all-into-conspiracy-theorists/. Accessed 6 Aug. 2017.

2. Doc, Ken. "MICK WEST." INVESTIGATE 9/11, 12 Sept. 2016, https://kendoc911 .wordpress.com/911-shills/mick-west/. Accessed 30 Nov. 2017.

3. Doc, Ken. "INVESTIGATE 9/11." https://kendoc911.wordpress.com/. Accessed 30 Nov. 2017.

4. Barkun, Michael. A Culture of Conspiracy: Apocalyptic Visions in Contemporary America. 2003, University of California Press.

5. Levin, Nicholas. "WHAT IS YOUR HOP LEVEL? Ten 9/11 Paradigms [summeroftruth.org]." Web.archive.org, 1 Apr. 2004. https://web.archive.org/web /20041030111442/http://summeroftruth.org/lihopmihopnohop.html. Accessed 7 Aug. 2017.

Chapter 3

1. Reddit User, "DebateFlatEarth." 2017, https://goo.gl/WwdLXC.

2. Fleming, Michael David. "Mick West: Busted Internet Troll at Metabunk." https:// www.michaeldavidfleming.com/mick-west-busted-internet-troll-at-metabunk/. Accessed 2 Jan. 2018.

3. Doc, Ken. "MICK WEST." Ken Doc—INVESTIGATE 9/11, 12 Sept. 2016, https:// kendoc911.wordpress.com/911-shills/mick-west/. Accessed 2 Jan. 2018.

4. West, Mick. "The IT2010 Competition." Mickwest.com, 11 Jan. 2007, http://mickwest .com/2007/01/11/the-it2010-competition/. Accessed 21 Nov. 2017.

5. American Red Cross. "Blood Donor Eligibility: Blood Pressure, Pregnancy, Disease & More." http://www.redcrossblood.org/donating-blood/eligibility-requirements /eligibility-criteria-alphabetical-listing. Accessed 8 Nov. 2017.

Chapter 4

1. Shuck, John. "The Science Behind the Collapse of the Three World Trade Center Towers on 9/11." Beta.prx.org, 9 Jun. 2017, https://beta.prx.org/stories/214113. Accessed 4 Sept. 2017.

2. Beard, Martin M. "What Happened?—My Story." Metabunk, 30 Mar. 2016, https://www.metabunk.org/martin-m-beard-what-happened-my-story.t7452/. Accessed 7 Nov. 2017.

3. WeAreChange Los Angeles. "9-11 Truth March 10-11-08 Santa Monica Part 3 FIREFIGHTERS—YouTube." Youtube.com, https://www.youtube.com/watch?v=E9lq6P9RrOg. Accessed 12 Nov. 2017.

4. Ventura, Jess. "Abby Martin Uncovers 9/11 | Jesse Ventura Off The Grid." Youtube.com, 12 Sept 2014, https://www.youtube.com/watch?v=6yi7XMrlEiU. Accessed 12 Nov. 2017.

5. Ventura, Jesse. "Counter-Conspiracy: World Trade Center 7 | Jesse Ventura Off The Grid." Youtube.com, https://www.youtube.com/watch?v=O3IvRPGVFmk. Accessed 12 Nov. 2017.

6. Lawson, Anthony. "WTC7—This is an Orange—YouTube." Youtube.com, https://www.youtube.com/watch?v=Zv7BImVvEyk. Accessed 12 Nov. 2017.

7. Benson, Buster. "Cognitive bias cheat sheet—Better Humans." Better Humans, 1 Sept. 2016, https://betterhumans.coach.me/cognitive-bias-cheat-sheet-55a472476b18. Accessed 20 Oct. 2017.

8. Dolan, Eric W. "Studies find the need to feel unique is linked to belief in conspiracy theories." PsyPost, 8 Aug. 2017, http://www.psypost.org/2017/08/studies-find-need-feel-unique-linked-belief-conspiracy-theories-49444. Accessed 20 Oct. 2017.

9. Sloat, Sarah. "Conspiracy Theorists Have a Fundamental Cognitive Problem, Say Scientists." Inverse, 17 Oct. 2017, https://www.inverse.com/article/37463-conspiracy-beliefs-illusory-pattern-perception. Accessed 20 Oct. 2017.

10. Dolan, Eric W. "Losers are more likely to believe in conspiracy theories, study finds." PsyPost, 17 Sept. 2017, http://www.psypost.org/2017/09/losers-likely-believe-conspiracy-theories-study-finds-49694. Accessed 20 Oct. 2017.

11. Bolton, Doug. "Conspiracy theorists are more likely to be suffering from stress, study finds." The Independent, 11 May 2016, http://www.independent.co.uk/news/science/conspiracy-theories-stress-psychology-study-anglia-ruskin-a7023966.html. Accessed 20 Oct. 2017.

12. Levesque, Danielle. "Narcissism and low self-esteem predict conspiracy beliefs." PsyPost, 25 Feb. 2016, http://www.psypost.org/2016/02/narcissism-and-low-self-esteem-predict-conspiracy-beliefs-41253. Accessed 20 Oct. 2017.

13. Zareva, Teodora. "Conspiracy Theories: Why the More Educated Don't Believe Them." Big Think, 12 Apr. 2017, http://bigthink.com/design-for-good/how-likely-are-you-to-believe-in-conspiracy-theories-depends-on-these-factors. Accessed 29 Dec. 2017.

14. Dolan, Eric W. "Study: The personal need to eliminate uncertainty predicts belief in conspiracy theories." PsyPost, 28 Jun. 2017, http://www.psypost.org/2017/06/study-personal-need-eliminate-uncertainty-predicts-belief-conspiracy-theories-49211. Accessed 20 Oct. 2017.

15. Rivas, Anthony. "Bipartisan Anti-Vaxxers And The Weird Science Behind Conspiracies." Medical Daily, 20 Jan. 2015, http://www.medicaldaily.com/conspiracy-theories-mostly-believed-people-far-left-right-political-spectrum-318502. Accessed 20 Oct. 2017.

16. To get this percentage you square the r value and multiply by 100.

17. IFLScience. "People Who Believe Conspiracy Theories Just Want To Be Unique, Say Psychologists." 18 Aug. 2017, http://www.iflscience.com/brain/people-who-believe-conspiracy-theories-just-want-to-be-unique-say-psychologists/. Accessed 25 Aug. 2017.

18. Dolan, Eric W. "Studies find the need to feel unique is linked to belief in conspiracy theories." PsyPost, 8 Aug. 2017, http://www.psypost.org/2017/08/studies-find-need-feel-unique-linked-belief-conspiracy-theories-49444. Accessed 25 Aug. 2017.

19. Edelson, Jack; Alduncin, Alexander; Sieja, James; & Uscinski, Joseph. "The Effect of Conspiratorial Thinking and Motivated Reasoning on Belief in Election Fraud." Political Research Quarterly 2017. http://journals.sagepub.com/doi/abs/10.1177/1065912917721061. Accessed 21 Sept. 2017.

20. Uscinski, Joseph; Parent, Joseph; & Torres, Bethany.. "Conspiracy Theories are for Losers." Papers.ssrn.com, 1 Aug. 2011, https://papers.ssrn.com/sol3/papers.cfm?abstract_id=1901755. Accessed 22 Sept. 2017.

21. De Keersmaecker, Jonas; & Roets, Arne. "'Fake news': Incorrect, but hard to correct. The role of cognitive ability on the impact of false information on social impressions." Sciencedirect.com, 7 Nov. 2017, https://www.sciencedirect.com/science/article/pii/S0160289617301617. Accessed 17 Feb. 2018.

22. Cichocka, Aleksandra; Marchlewska, Marta; & Golec de Zavala, Agnieszka. "Does Self-Love or Self-Hate Predict Conspiracy Beliefs? Narcissism, Self-Esteem, and the Endorsement of Conspiracy Theories." 2016, Journals.sagepub.com, http://journals.sagepub.com/doi/abs/10.1177/1948550615616170. Accessed 18 Feb. 2018.

23. Sunstein, Cass; & Vermeule, Adrian. "Conspiracy Theories." John M. Olin Program in Law and Economics Working Paper No. 387, 2008, http://chicagounbound.uchicago.edu/cgi/viewcontent.cgi?article=1118&context=law_and_economics.

24. "Conspiracy Road Trip: UFOs." BBC, Oct 15, 2012, https://youtube/7ByWCFX4ZQs?t=53m37s. Accessed 31 Aug. 2017.

Chapter 5

1. Cairns, Rose. "Climates of suspicion: 'Chemtrail' conspiracy narratives and the international politics of geoengineering." Onlinelibrary.wiley.com, 25 Nov. 2014, http://onlinelibrary.wiley.com/doi/10.1111/geoj.12116/full. Accessed 7 Jan. 2018.

2. Sagan, Carl. Demon-Haunted World: Science as a Candle in the Dark (Kindle Locations 7218-7220). Random House Publishing Group. Kindle Edition.

3. Dunne, Carey. "My month with Chemtrails conspiracy theorists." the Guardian, 22 May 2017, http://www.theguardian.com/environment/2017/may/22/california -conspiracy-theorist-farmers-Chemtrails. Accessed 10 Aug. 2017.

4. Nyhan, Brendan; & Reifler, Jason. "When Corrections Fail: The Persistence of Political Misperceptions." SpringerLink, 1 Jun. 2010, https://link.springer.com/article /10.1007/s11109-010-9112-2. Accessed 1 Feb. 2018.

5. Cook, John; & Lewandowsky, Stephan. *The Debunking Handbook*. Skepticalscience. com, 22 Jan. 2012, https://www.skepticalscience.com/docs/Debunking_Handbook .pdf. Accessed 1 Feb. 2018.

6. Wood, Thomas; & Porter, Ethan. "The Elusive Backfire Effect: Mass Attitudes' Steadfast Factual Adherence." Papers.ssrn.com, 31 Dec. 2017, https://papers.ssrn .com/sol3/papers.cfm?abstract_id=2819073. Accessed 1 Feb. 2018.

7. Chan, Man-pui Sally, et al. "Debunking: A Meta-Analysis of the Psychological Efficacy of Messages Countering Misinformation.—PubMed—NCBI." Ncbi.nlm.nih.gov, https://www.ncbi.nlm.nih.gov/pubmed/28895452. Accessed 1 Feb. 2018.

8. West, Mick. "Debunked: The 'short-lived' fires of WTC 1 & 2." Metabunk, 12 Jun. 2013, https://www.metabunk.org/debunked-the-short-lived-fires-of-wtc-1-2.t1771/. Accessed 17 Feb. 2018.

9. The Associated Press. "Landfill burned for weeks; property owner facing charges." Sun-Sentinel.com, 5 Dec. 2017, http://www.sun-sentinel.com/local/miami-dade/fl -reg-dade-illegal-landfill-20171205-story.html. Accessed 17 Feb. 2018.

10. Foran, Chris. " 'Like looking down on hell': When a Milwaukee dump caught fire and burned for months." Milwaukee Journal Sentinel, 14 Nov. 2017, https://www .jsonline.com/story/life/green-sheet/2017/11/14/like-looking-down-hell-when -milwaukee-dump-caught-fire-and-burned-months/856767001/. Accessed 17 Feb. 2018.

11. Tribune Wire. "Underground fire outside St. Louis has burned since 2010, nears nuclear waste dump." chicagotribune.com, 10 Oct. 2015, http://www.chicagotribune .com/news/nationworld/midwest/ct-st-louis-underground-fire-20151010-story.html. Accessed 17 Feb. 2018.

Chapter 6

1. Wikipedia: "McVeigh felt the need to personally reconnoiter sites of rumored conspiracies. He visited Area 51 in order to defy government restrictions on photography and went to Gulfport, Mississippi to determine the veracity of rumors about United Nations operations." https://en.wikipedia.org/wiki/Timothy_McVeigh.

Chapter 7

1. West, Mick. "A brief history of 'Chemtrails'—Contrail Science." Contrail Science, 11 May 2007, http://contrailscience.com/a-brief-history-of-Chemtrails/. Accessed 30 Oct. 2017.

2. Izrael, Yuri, et al. "Field experiment on studying solar radiation passing through aerosol layer." SpringerLink, 1 May 2009, https://link.springer.com/article/10.3103 /S106837390905001X. Accessed 30 Oct. 2017.

3. Russell, Lynn M., et al., "Eastern Pacific Emitted Aerosol Cloud Experiment (E-PEACE)." Research Profiles, http://scrippsscholars.ucsd.edu/gcroberts/content /eastern-pacific-emitted-aerosol-cloud-experiment-e-peace. Accessed 30 Oct. 2017.

4. Keith, David W.; Duren, Riley; & MacMartin, Douglas G. "Field experiments on solar geoengineering: report of a workshop exploring a representative research portfolio." PubMed Central (PMC), https://www.ncbi.nlm.nih.gov/pmc/articles/PMC4240958/. Accessed 30 Oct. 2017.

5. Dykema, John A. "Stratospheric controlled perturbation experiment: a small-scale experiment to improve understanding of the risks of solar geoengineering." Philosophical Transactions of the Royal Society of London A: Mathematical, Physical and Engineering Sciences, 28 Dec. 2014, http://rsta.royalsocietypublishing.org/content /372/2031/20140059. Accessed 30 Oct. 2017.

6. Temple, James. "Harvard scientists are gearing up for some of the first outdoor geo-engineering experiments." MIT Technology Review, 29 Mar. 2017, https://www .technologyreview.com/s/603974/harvard-scientists-moving-ahead-on-plans-for -atmospheric-geoengineering-experiments/. Accessed 30 Oct. 2017.

7. MacMartin, Douglas G. "Geoengineering with stratospheric aerosols: What do we not know after a decade of research?" Onlinelibrary.wiley.com, 18 Nov. 2016, http:// onlinelibrary.wiley.com/doi/10.1002/2016EF000418/full. Accessed 15 Feb. 2018.

8. West, Mick. "Calculator for RHi and Contrail Persistence Criteria." Metabunk, 12 Jan. 2016, https://www.metabunk.org/calculator-for-rhi-and-contrail-persistence -criteria.t7196/. Accessed 1 Nov. 2017.

9. American Meteorological Society. "Mixing cloud—AMS Glossary." Glossary.amet-soc.org, 11 Jul. 2016, http://glossary.ametsoc.org/wiki/Mixing_cloud. Accessed 1 Nov. 2017.

10. Thomas, William. "Frequently Aired Criticisms." Web.archive.org, https://web .archive.org/web/19990508091715/http://www.islandnet.com:80/~wilco /Chemtrailsfacs.htm. Accessed 19 Feb. 2018.

11. West, Mick. "Pre 1995 Persistent Contrail Archive." Metabunk, 23 Mar. 2012, https:// www.metabunk.org/pre-1995-persistent-contrail-archive.t487/. Accessed 31 Oct. 2017.

12. West, Mick. "Debunking 'Contrails don't persist' with 70 years of books on clouds." YouTube, 28 Feb, 2014, https://www.youtube.com/watch?v=X72uACIN_00. Accessed 31 Oct. 2017.

13. Wigington, Dane. "High Bypass Turbofan Jet Engines, Geoengineering, And The Contrail Lie." Geoengineering Watch, 9 Sep. 2017, http://www.geoengineeringwatch.org/the-contrail-lie/. Accessed 1 Nov. 2017.

14. ibid.

15. West, Mick. "Debunked: High Bypass Turbofans do not make Contrails [actually they make more]." Metabunk, 26 Feb. 2014, https://www.metabunk.org/debunked-high-bypass-turbofans-do-not-make-contrails-actually-they-make-more.t3187/. Accessed 1 Nov. 2017.

16. Marcianò, Rosario / Tanker Enemy. "The insider: Chemtrails KC-10 sprayer air to air—The proof." https://www.youtube.com/watch?v=bSSWnXQsgOU. Accessed 29 Nov. 2017.

17. M., Tim. "The original 'KC-10 spreading Chemtrails' spoof video." Mar 12 2011, https://www.youtube.com/watch?v=t22wy4c-A-A. Accessed 29 Nov. 2017.

18. M., Tim. "KC-10 Still pics." https://www.youtube.com/watch?v=SGNyZ9PKtyI. Accessed 29 Nov. 2017.

19. Geoengineering Watch. "Smoking Gun Proof Of Atmospheric Spraying." Dec 2 2014, http://www.geoengineeringwatch.org/smoking-gun-proof-of-atmospheric-spraying/. Accessed 29 Nov. 2017.

20. Trans-Pecos WMA. "Seeding Report." Wtwma.com, 24 Aug. 2017, http://wtwma.com/Daily%20Operations/TPWMA/08242017T.pdf. Accessed 31 Aug. 2017.

21. Dick Van Dyke Show. "Cloud Seeding Mentioned on the Dick Van Dyke Show, 1965." https://www.youtube.com/watch?v=5VUaHvTiwkg. Accessed 27 Nov. 2017.

22. U.S. Air Service. Reprinted in Monthly Weather Review, July 1921, http://contrailscience.com/files/mwr-049-07-0412c.pdf.

23. Yenne, Bill. "Inside Boeing: Building the 777." Page 76.

24. Boeing. "Boeing 747-100 First Flight 2/2." https://youtu.be/xrDv4jX_MUs?t=58s. Accessed 29 Nov. 2017.

25. West, Mick. "'Chemtrail' Aircraft Photos." Contrail Science, 23 May 2007, http://contrailscience.com/contrail-or-Chemtrail/. Accessed 29 Nov. 2017.

26. Legal Alliance to Stop Geoengineering. "Re: Notice of Intent to File Citizens' Suits Pursuant to Federal Clean Water Act and Federal Safe Drinking Water Act." Geoengineeringwatch.org, 25 Jul. 2016, http://www.geoengineeringwatch.org/documents/LASG%2060-Day%20Notice%20-%20web%20version.pdf. Accessed 29 Nov. 2017.

27. USGS. "Geochemistry & Mineralogy of U.S. Soils." https://mrdata.usgs.gov/soilgeochemistry/#/detail/element/13. Accessed 13 Feb. 2018.

28. CDC. "Aluminum Toxicological Profile." 16 Apr. 2013, https://www.atsdr.cdc.gov/toxprofiles/tp22-c1.pdf. Accessed 29 Nov. 2017.

29. Ibid.

30. "How to Test WATER for Aerosol Spraying Evidence." Geoengineeringwatch.org, 31 Jul. 2013, http://www.geoengineeringwatch.org/html/watertesting.html. Accessed 29 Nov. 2017.

31. West, Mick. "Wigington/West Geoengineering Debate." Metabunk, 14 Aug. 2013, https://www.metabunk.org/wigington-west-geoengineering-debate.t2211/. Accessed 15 Feb. 2018.

32. West, Mick. "Debunked: Shasta Snow and Water Aluminum Tests." Metabunk, 18 Apr. 2011, https://www.metabunk.org/posts/616/. Accessed 15 Feb. 2018.

33. Geoengineering Watch. "Extensive List of PATENTS." Aug 8, 2012, Geoengineering Watch, http://www.geoengineeringwatch.org/an-extensive-list-of-patents/. Accessed 6 Dec. 2017.

34. West, Mick. "Debunked: Patents. As Evidence of Chemtrails, Geoengineering, Existence, Operability, or Intent." Metabunk, 24 May 2014, https://www.metabunk.org/debunked-patents-as-evidence-of-Chemtrails-geoengineering-existence-operability-or-intent.t3704/. Accessed 15 Feb. 2018.

35. "USPTO Museum Opens New Exhibit Showcasing American Ingenuity." Uspto.gov, 11 Feb. 2002, https://www.uspto.gov/about-us/news-updates/uspto-museum-opens-new-exhibit-showcasing-american-ingenuity. Accessed 6 Dec. 2017.

36. LII / Legal Information Institute. "Patent." LII / Legal Information Institute, 6 Aug. 2007, https://www.law.cornell.edu/wex/patent. Accessed 6 Dec. 2017.

37. International Space Elevator Consortium, https://isec.org/. Accessed 18 Mar. 2018.

38. Walker, Jay. "Our System Is So Broken, Almost No Patented Discoveries Ever Get Used." WIRED, 5 Jan. 2015, https://www.wired.com/2015/01/fixing-broken-patent-system/. Accessed 6 Dec. 2017.

39. Wang, J., et al. "Performance of operational radiosonde humidity sensors." Radiometrics.com, 23 Aug 2003, http://radiometrics.com/data/uploads/2012/11/Refsonde-GRL.pdf. Accessed 19 Feb. 2018.

40. West, Mick. "Relative Humidity at Flight Altitudes. Resources and Contrail prediction." Metabunk, 22 Oct. 2016, https://www.metabunk.org/relative-humidity-at-flight-altitudes-resources-and-contrail-prediction.t8084/. Accessed 19 Feb. 2018.

41. Agorist, Matt. "No Longer Conspiracy: CIA Admits Plans Of Aerosol Spraying For Geoengineering." MintPress News, 7 Jul. 2016, http://www.mintpressnews.com/no-longer-conspiracy-cia-admits-plans-aerosol-spraying-geoengineering/218179/. Accessed 16 Jan. 2018.

42. Brennan, John O. "Director Brennan Speaks at the Council on Foreign Relations—Central Intelligence Agency." Cia.gov, 19 Jan. 2017, https://www.cia.gov/news-information/speeches-testimony/2016-speeches-testimony/director-brennan-speaks-at-the-council-on-foreign-relations.html. Accessed 16 Jan. 2018.

Chapter 8

1. Wittschier, Stephanie. "Die lockere Schraube." Dielockereschraube.de, 4 Nov. 2017, http://www.dielockereschraube.de/. Accessed 9 Dec. 2017.

2. "Brigitta Zuber—Chemtrails und HAARP." https://www.youtube.com/watch?v
   =fY82EBzPGEs. Accessed 9 Dec. 2017.

Chapter 9

1. Sure, Fran. "Psychology Experts Speak Out: Why is the 9/11 Evidence Difficult for
   Some to Accept?" http://www1.ae911truth.org/home/645-psychology-experts-speak
   -out-why-is-the-911-evidence-difficult-for-some-to-accept-.html. Accessed 15 Nov.
   2017.
2. Facebook comment. Nov 17 2017, https://goo.gl/iKogPW.
3. "AE911Truth—Architects & Engineers Investigating the destruction of all three
   World Trade Center skyscrapers on September 11." http://www.ae911truth.org/.
   Accessed 15 Nov. 2017.
4. Gage, Richard. "Why are Architects and Engineers Re-examining the WTC Collapses?" AE911Truth.org, https://web.archive.org/web/20070817205947/http://www
   .ae911truth.org:80/info/4. Accessed 19 Nov. 2017.
5. Hoffman, Jim. "Hypothetical Blasting Scenario." 911research.wtc7.net, 14 Feb. 2010,
   http://911research.wtc7.net/essays/thermite/blasting_scenario.html. Accessed 25 Jan.
   2018.
6. http://www.ae911truth.org/signatures/ae.html.
7. West, Mick. "Debunked: AE911Truth's Analysis of Slag Residue from WTC Debris."
   Metabunk, 27 Jan. 2018, https://www.metabunk.org/debunked-ae911truths-analysis
   -of-slag-residue-from-wtc-debris.t9468/. Accessed 1 Feb. 2018.
8. West, Mick. "Debunked: The WTC 9/11 Angle Cut Column. [Not Thermite, Cut
   Later]." Metabunk, 27 Jan. 2018, https://www.metabunk.org/debunked-the-wtc-9-11
   -angle-cut-column-not-thermite-cut-later.t9469/. Accessed 1 Feb. 2018.
9. Chandler, David. "Why I Am Convinced 9/11 Was an Inside Job." Apr. 20 2014,
   http://www1.ae911truth.org/en/news-section/41-articles/874-. Accessed 16 Feb. 2018.
10. West, Mick. "Molten Steel in the Debris Pile, Cool Down Time?" Metabunk, 18 Nov.
    2017, https://www.metabunk.org/molten-steel-in-the-debris-pile-cool-down-time
    .t9255/. Accessed 16 Feb. 2018.
11. Rivera, Michael. "The World Trade Center Demolition: An Analysis." Before Jun 22
    2003, https://web.archive.org/web/20030622150344/http://whatreallyhappened.com:
    80/shake2.html. Accessed 16 Feb. 2018.
12. Rivera, Michael. "Ron Brown: Evidence Of A Cover up | WHAT REALLY HAP-
    PENED." Whatreallyhappened.com, http://www.whatreallyhappened.com
    /RANCHO/CRASH/BROWN/brown.php. Accessed 16 Feb. 2018.
13. Mythbusters. "Episode 113—End With a Bang." Nov 12 2008.
14. Lee, R. J. Correspondence with Ron Wieck, Nmsr.org, Feb 2012, http://www.nmsr
    .org/nmsr911.htm. Accessed 16 Feb. 2018.

15. West, Mick. "Making Iron Microspheres—Grinding, Impacts, Welding, Burning." Metabunk, 2 Feb. 2018, https://www.metabunk.org/making-iron-microspheres-grinding-impacts-welding-burning.t9533/. Accessed 18 Mar. 2018.

16. West, Mick. "Investigating 'Active Thermitic Material Discovered in Dust from the 9/11 WTC Catastrophe.'" Metabunk, 2 Feb. 2018, https://www.metabunk.org/investigating-active-thermitic-material-discovered-in-dust-from-the-9-11-wtc-catastrophe.t9485/. Accessed 16 Feb. 2018.

17. "World Trade Center Building 7 Demolished on 9/11? | AE911Truth." http://www1.ae911truth.org/. Accessed 23 Jan. 2018.

18. "Dutch tv show Zembla investigates 911 theories—2006 (English subs)—Video Dailymotion." Dailymotion, 16 Apr. 2016, http://www.dailymotion.com/video/x44yy1c. Accessed 1 Feb. 2018.

19. "Danny Jowenko's old interview about the Twin Towers wtc1 & wtc2." https://www.youtube.com/watch?v=k3wwdI0XawI&t=151s. Accessed 1 Feb. 2018.

20. Newspolls.org. "Question/VAR 28." https://web.archive.org/web/20060806155035/http://www.newspolls.org/question.php?question_id=717. Accessed 22 Jan. 2018.

21. Mckee, Craig. "Stand with us: add your name to the 'No 757 hit the Pentagon on 9/11' list!" Truth and Shadows, 11 Dec. 2017, https://truthandshadows.wordpress.com/2017/12/11/the-no-757-hit-the-pentagon-list/. Accessed 23 Jan. 2018.

22. "Meet the AE911Truth Board of Directors." Architects and Engineers for 9/11 Truth, http://www1.ae911truth.org/home/678-meet-the-ae911truth-board-of-directors.html. Accessed 23 Jan. 2018.

23. Jenkins, Ken. "The Pentagon Plane Puzzle (preview—ver 3)." Youtube.com, https://www.youtube.com/watch?v=YsadQzNhT-Q&feature=youtu.be. Accessed 14 Feb. 2018.

24. Mick West, et al. "Does this photo show a too-small hole in the Pentagon? [No]." Metabunk, 5 Jan. 2017, https://www.metabunk.org/does-this-photo-show-a-too-small-hole-in-the-pentagon-no.t8302/. Accessed 23 Jan. 2018.

25. Ibid.

26. "Inexperienced Pilot Recreating 9/11 Flight 77's Descending Turn into the Pentagon." Youtube.com, https://www.youtube.com/watch?v=UaOLpeTC7hY&lc=Ugw3Mvs1K4cI05CiwaR4AaABAg. Accessed 23 Jan. 2018.

27. Mlakar, Paul E., et al. *The Pentagon Building Performance Report.* Ascelibrary.org, https://ascelibrary.org/doi/book/10.1061/9780784406380. Accessed 23 Jan. 2018.

28. Ashley, Victoria, et al. "The Pentagon Event." Scholars for 9/11 Truth, 2 May 2016, http://www.scientificmethod911.org/docs/Honegger_Hypothesis_042916.pdf. Accessed 15 Feb. 2018.

29. Griffin, David Ray. *The New Pearl Harbor Revisited.* 2008, Olive Branch Press. Page 104.

30. DoD. "Transforming Department of Defense Financial Management." defense.gov, 10 Jul. 2001, http://archive.defense.gov/news/Jul2001/d20010710finmngt.pdf. Accessed 23 Jan. 2018.

31. RT. "Black Budget: US govt clueless about missing Pentagon $trillions." https://www .youtube.com/watch?v=j4dzECaBxFU. Accessed 18 Feb. 2018.

32. RT. "$21 trillion of unauthorized spending by US govt discovered by economics professor." RT International, 16 Dec. 2017, https://www.rt.com/usa/413411-trillions -dollars-missing-research/. Accessed 18 Feb. 2018.

33. NIST. "Final Reports from the NIST World Trade Center Disaster Investigation." 27 Jun. 2012, https://www.nist.gov/engineering-laboratory/final-reports-nist-world-trade -center-disaster-investigation. Accessed 19 Feb. 2018.

34. Mcallister, Terese P., et al. "Structural Fire Response and Probable Collapse Sequence of World Trade Center Building 7, Federal Building and Fire Safety Investigation of the World Trade Center Disaster (NIST NCSTAR 1-9) VOLUMES 1 and 2." NIST, 20 Nov 2008, http://ws680.nist.gov/publication/get_pdf.cfm?pub_id=861611 Accessed 19 Feb. 2018.

35. NIST. "FAQs—NIST WTC Towers Investigation." NIST, 21 Sept. 2016, https://www .nist.gov/el/faqs-nist-wtc-towers-investigation. Accessed 19 Feb. 2018.

36. NIST. "FAQs—NIST WTC 7 Investigation." NIST, 21 Sept. 2016, https://www.nist.gov /el/faqs-nist-wtc-7-investigation. Accessed 19 Feb. 2018.

37. "Vérinage Compilation—Explosiveless Demolition." https://www.youtube.com /watch?v=aYuBdR3CvY4. Accessed 19 Feb. 2018.

38. "Sounds of Demolition." Youtube.com, https://www.youtube.com/watch?v =vMpJQgbKsms. Accessed 19 Feb. 2018.

39. "9/11: WTC 7 Collapse (NIST FOIA, CBS video)." https://www.youtube.com/watch?v =nqbUkThGlCo. Accessed 19 Feb. 2018.

40. "New Detail Surface After Two Major Rock Slides At Yosemite." https://youtu.be /T-l5flDozts?t=19s Accessed 19 Feb. 2018.

41. purgatoryironworks. "For the undying 9/11 MORONIC JET FUEL ARGUMENT." https://www.youtube.com/watch?v=FzF1KySHmUA. Accessed 19 Feb. 2018.

42. National Geographic, "Evidence that steel can be weakened by a jet fuel fire." https:// www.youtube.com/watch?v=N2TMVDYpp2Q&t=270s. Accessed 19 Feb. 2018.

43. Ibid.

44. "World Trade Center—Role of floor loss and buckling." https://www.youtube.com /watch?v=VGhTTUBuMYo. Accessed 19 Feb. 2018.

45. West, Mick. "Static Force vs. Dynamic force." https://www.youtube.com/watch?v =wZCFo3Lcbx8. Accessed 19 Feb. 2018.

Endnotes

Chapter 10

1. Current, Edward, "Building 7 Explained: The Tube That Crumpled." 2011, Updated 2017. https://www.youtube.com/watch?v=4LUDXpMhkNk. Accessed 19 Feb. 2018.
2. Kruger, J.; & Dunning, D. "Unskilled and unaware of it: how difficulties in recognizing one's own incompetence lead to inflated self-assessments.—PubMed—NCBI." Ncbi.nlm.nih.gov, https://www.ncbi.nlm.nih.gov/pubmed/10626367. Accessed 18 Mar. 2018.

Chapter 11

1. DoD memo. "Justification for US Military Intervention in Cuba." 13 Mar. 1962, Page 5, http://nsarchive.gwu.edu/news/20010430/index.html. Accessed 3 Aug. 2017.
2. Bamford, James. *Body of Secrets: Anatomy of the Ultra Secret National Security Agency.* New York: Anchor Books, 2001.
3. YouTube. "Jesse Ventura and Judge Napolitano: Operation Northwoods, 9/11, and Wikileaks." 6 Apr. 2011, https://www.youtube.com/watch?v=8sG9TaziF_Q. 1m48s. Accessed 3 Aug. 2017.
4. YouTube. "Operation Northwoods Exposed (MUST-SEE VIDEO!!)." 10 Dec. 2008, https://www.youtube.com/watch?v=Rp3P2wDKQK4. Accessed 3 Aug. 2017. (Video Clip Compilation of Ventura, Bamford and Jones on Operation Northwoods).
5. "Operation Northwoods Documents." Nsarchive.gwu.edu, 1962, http://nsarchive.gwu.edu/news/20010430/northwoods.pdf. Accessed 2 Aug. 2017.
6. "Foreign Relations of the United States, 1961–1963, Volume X, Cuba, January 1961–September 1962—Office of the Historian." 30 May 2017, https://history.state.gov/historicaldocuments/frus1961-63v10/comp1. Accessed 4 Aug. 2017.
7. "Lyndon B. Johnson: Radio and Television Report to the American People Following Renewed Aggression in the Gulf of Tonkin." 4 Aug 1964, http://www.presidency.ucsb.edu/ws/index.php?pid=26418&st=destroyer+Maddox&st1=. Accessed 4 Aug. 2017.
8. "Newly Declassified National Security Agency History Questions Early Vietnam War Communications Intelligence." Nsarchive2.gwu.edu, 22 Sept. 2017, http://nsarchive2.gwu.edu//NSAEBB/NSAEBB132/press20051201.htm. Accessed 9 Oct. 2017.
9. Hanyok, Robert J. "Skunks, Bogies, Silent Hounds, and the Flying Fish: The Gulf of Tonkin Mystery, 2–4 August 1964." Cryptologic Quarterly, 10 Aug. 2017, http://nsarchive2.gwu.edu//NSAEBB/NSAEBB132/relea00012.pdf. Accessed 9 Oct. 2017.
10. "LBJ Tapes on the Gulf of Tonkin Incident." Nsarchive2.gwu.edu, 22 Sept. 2017, http://nsarchive2.gwu.edu//NSAEBB/NSAEBB132/tapes.htm. Accessed 9 Oct. 2017.
11. Operation Northwoods Documents, Page 7.
12. Various. "Debunked: Sandy Hook: The Man in The Woods." Metabunk, 19 Jan. 2013, https://www.metabunk.org/debunked-sandy-hook-the-man-in-the-woods.t1077/. Accessed 6 Feb. 2018.

13. Qiu, Linda. "No, There Was Not More Than One Gunman in the Las Vegas Shooting." Nytimes.com, 4 Oct. 2017, https://www.nytimes.com/2017/10/04/us/politics/fact-check-vegas-gunman.html. Accessed 6 Feb. 2018.

14. Evon, Dan. "'Multiple Shooters' at Orlando Nightclub." Snopes.com, 13 June 2016, https://www.snopes.com/multiple-shooters-orlando-cody-agnew/. Accessed 6 Feb. 2018.

15. Howe, M. L. "The Neuroscience of Memory: Implications for the Courtroom." PubMed Central (PMC), https://www.ncbi.nlm.nih.gov/pmc/articles/PMC4183265/. Accessed 6 Feb. 2018.

16. Mendoza, Marilyn. "The Healing Power of Laughter in Death and Grief." Psychology Today, https://www.psychologytoday.com/blog/understanding-grief/201611/the-healing-power-laughter-in-death-and-grief. Accessed 7 Apr. 2018.

Chapter 12

1. "Jet fuel can't melt steel beams" is a common refrain of 9/11 Truthers. They don't realize that it can weaken them enough so they bend and fail.

2. purgatoryironworks. "For the undying 9/11 MORONIC JET FUEL ARGUMENT." Youtube.com, https://www.youtube.com/watch?v=FzF1KySHmUA. Accessed 15 Feb. 2018.

Chapter 13

1. Garwood, Christine. *Flat Earth: The History of an Infamous Idea* (pp. 343–344). 2008, St. Martin's Press. Kindle Edition.

2. "The Flat Earth Society." https://theflatearthsociety.org/. Accessed 8 Jan. 2018.

3. "Eddie Bravo Flat Earth Debate with Earth Scientist." https://www.youtube.com/watch?v=cKuhulvuEsY. Accessed 10 Jan. 2018.

4. Robotham, Samuel. *Zetetic Astronomy: Earth Not a Globe!* Google Books, https://books.google.com/books/about/Zetetic_astronomy_Earth_not_a_globe_an_e.html?id=oTUDAAAAQAAJ. Accessed 10 Jan. 2018.

5. Dubay, Eric. *The Flat Earth Conspiracy.* 2014, p. 135. Lulu.com. Kindle Edition.

6. By Pythagoras's theorem: 4,000 miles is the base of a triangle from you to under the Sun. 3,000 miles is the height. So the distance from you to the Sun is the hypotenuse. $3^2 + 4^2 = 5^2$, so it's 5,000 miles away.

7. West, Mick. "How to Verify that the Sun is a Distant Sphere, with Binoculars & Sunspots." Metabunk, 28 Apr. 2017, https://www.metabunk.org/how-to-verify-that-the-sun-is-a-distant-sphere-with-binoculars-sunspots.t8646/. Accessed 25 Jan. 2018.

Chapter 15

1. Parker, Ryan. "Charlie Sheen on His Surprising New '9/11' Movie and Those Truther Comments." The Hollywood Reporter, 8 Sept. 2017, http://www.hollywoodreporter.com/news/charlie-sheen-9-11-movie-truther-comments-1036494. Accessed 11 Sept. 2017.

2. "Feynman: Magnets FUN TO IMAGINE 4—YouTube." Youtube.com, https://www .youtube.com/watch?v=wMFPe-DwULM. Accessed 19 Feb. 2018.

3. Gage, Richard. "For the undying 9/11 MORONIC STEEL = AIR ARGUMENT." https://www.youtube.com/watch?v=FvuKUmK9eB0. Accessed 7 Apr. 2018.

4. Taylor, Jon. "MLM's ABYSMAL NUMBERS." Ftc.gov, 11 Jul. 2013, https://www.ftc .gov/sites/default/files/documents/public_comments/trade-regulation-rule -disclosure-requirements-and-prohibitions-concerning-business-opportunities-ftc .r511993-00008%C2%A0/00008-57281.pdf. Accessed 6 Dec. 2017.

5. Letaio, Mary. "Mysterious Skin Disease." Medhelp.org, 21 Aug. 2002, https://www .medhelp.org/posts/Dermatology/Mysterious-Skin-Disease/show/241305. Accessed 1 Feb. 2018.

6. Jines, Anda. "10 Tips for Helping a Loved One Cope with Illness Anxiety." Hoover. associates, https://hoover.associates/archives/3761. Accessed 1 Feb. 2018.

7. Van Noppen, Barbara. "Living With Someone Who Has OCD. Guidelines for Family Members." International OCD Foundation, 25 May 2014, https://iocdf.org/expert -opinions/expert-opinion-family-guidelines/. Accessed 1 Feb. 2018.

8. Sunstein, Cass; & Vermeule, Adrian. "Conspiracy Theories." Chicagounbound .uchicago.edu, https://chicagounbound.uchicago.edu/cgi/viewcontent.cgi. Accessed 13 Feb. 2018.

Chapter 16

1. Collins, Ben; & Cox, Joseph. "Jenna Abrams, Russia's Clown Troll Princess, Duped the Mainstream Media and the World." The Daily Beast, 3 Nov. 2017, https://www .thedailybeast.com/jenna-abrams-russias-clown-troll-princess-duped-the -mainstream-media-and-the-world. Accessed 12 Nov. 2017.

2. Kalugin, Oleg (interview). "CNN—Cold War Experience: Espionage." Web.archive .org, January 1998https://web.archive.org/web/20070206020316/http://www.cnn .com/SPECIALS/cold.war/episodes/21/interviews/kalugin/. Accessed 12 Nov. 2017.

3. RT International. "Americans continue to fight for 9/11 truth." RT International, https://www.rt.com/usa/americans-fight-911/. Accessed 13 Nov. 2017.

4. Ventura, Jesse. RT International. "For some, the search for what happened on 9/11 isn't over." RT International, https://www.rt.com/usa/jesse-ventura-911-truth/. Accessed 13 Nov. 2017.

5. Duff, Gordon. "America after Sandy Hook, Disarmed and Silenced." Veterans Today, 27 Jan. 2013, https://www.veteranstoday.com/2013/01/27/america-after-sandy-hook -disarmed-and-silenced/. Accessed 18 Feb. 2018.

6. Luna, Francisco. "Chemtrails y el Cambio Climático." RT en Español, 8 Dec. 2009, https://actualidad.rt.com/opinion/iluminando_conciencias/view/4639-Chemtrails-y -Cambio-Climatico. Accessed 18 Feb. 2018.

7. Gerrans, Sam. "YouTube and the art of investigation—RT Op-Edge." RT International, https://www.rt.com/op-edge/316642-youtube-art-investigation-research/. Accessed 13 Nov. 2017.

8. RT International. "RT watched by 70mn viewers weekly, half of them daily—Ipsos survey—RT World News." RT International, 10 Mar. 2016, https://www.rt.com/news/335123-rt-viewership-ipsos-study/. Accessed 3 Apr. 2018.

9. Norton. "What Is A Botnet?" 2 Apr. 2018, https://us.norton.com:80/internetsecurity-malware-what-is-a-botnet.html. Accessed 3 Apr. 2018.

10. Marciel, Miriam, et al. "Understanding the Detection of View Fraud in Video Content Portals." Arxiv.org, 7 Feb. 2016, https://arxiv.org/pdf/1507.08874.pdf. Accessed 3 Apr. 2018.

11. Gross, Grant. "Senator: Russia used 'thousands' of internet trolls during U.S. election." Computerworld, 30 Mar. 2017, https://www.computerworld.com/article/3186642/security/senator-russia-used-thousands-of-internet-trolls-during-us-election.html. Accessed 13 Nov. 2017.

12. BBC Trending In-depth Stories On Social Media. "'Russian trolls' promoted California independence." BBC News, http://www.bbc.com/news/blogs-trending-41853131. Accessed 13 Nov. 2017.

13. BBC News. "Russia posts 'reached 126m Facebook users.'" BBC News, http://www.bbc.com/news/world-us-canada-41812369. Accessed 13 Nov. 2017.

14. Tufekci, Zeynep. "We're building a dystopia just to make people click on ads | Zeynep Tufekci." Youtube.com, https://www.youtube.com/watch?v=iFTWM7HV2UI&feature=share. Accessed 26 Nov. 2017.

15. WIRED. "Google Is 2 Billion Lines of Code—And It's All in One Place." WIRED, 16 Sept. 2015, https://www.wired.com/2015/09/google-2-billion-lines-codeand-one-place/. Accessed 4 Apr. 2018.

16. Cuban, Mark. Twitter, 28 Jan. 2018, https://twitter.com/mcuban/status/957686987229618176. Accessed 7 Feb. 2018.

17. Spangler, Todd. "Mark Zuckerberg: 'Pretty Crazy Idea' That Facebook Fake News Helped Donald Trump Win." Variety, 11 Nov. 2016, http://variety.com/2016/digital/news/mark-zuckerberg-facebook-donald-trump-win-1201915811/. Accessed 16 Nov. 2017.

18. Zuckerberg, Mark. "I want to respond to President Trump's tweet. . . ." Facebook.com, https://www.facebook.com/zuck/posts/10104067130714241?pnref=story. Sept 27, 2017, Accessed 18 Feb. 2018.

19. Perlroth, Nicole; Frenkel, Sheera; & Shane, Scott. "Facebook Executive Planning to Leave Company Amid Disinformation Backlash." Nytimes.com, 19 Mar. 2018, https://www.nytimes.com/2018/03/19/technology/facebook-alex-stamos.html. Accessed 19 Mar. 2018.

20. Frier, Sarah. "Facebook Stumbles With Early Effort to Stamp Out Fake News." Bloomberg.com, 30 Oct. 2017, https://www.bloomberg.com/news/articles/2017-10-30 /facebook-stumbles-with-early-effort-to-stamp-out-fake-news. Accessed 16 Nov. 2017.

21. Smith, Jeff. "Designing Against Misinformation—Facebook Design—Medium." Medium, 20 Dec. 2017, https://medium.com/facebook-design/designing-against -misinformation-e5846b3aa1e2. Accessed 15 Mar. 2018.

22. Google. "Fact Check now available in Google Search and News around the world." Google, 7 Apr. 2017, https://www.blog.google/products/search/fact-check-now -available-google-search-and-news-around-world/. Accessed 7 Feb. 2018.

23. Mckay, Tom. "Conservatives Are Now Getting Angry About Google's Fact-Checking Module." Gizmodo, 9 Jan. 2018, https://gizmodo.com/conservatives-are-now-getting -angry-about-googles-fact-1821934885. Accessed 7 Feb. 2018.

24. Mckay, Tom. "Microsoft: Bing's US Market Share Is Wildly Underestimated." Gizmodo, 19 Aug. 2017, https://gizmodo.com/microsoft-bings-us-market-share-is-wildly -underestimat-1798053061. Accessed 18 Feb. 2018.

25. Intelligence and Autonomy. "About I&A." Intelligence and Autonomy, 21 Jul. 2015, https://autonomy.datasociety.net/about/. Accessed 7 Feb. 2018.

26. Santa Clara University. "The Trust Project—Markkula Center for Applied Ethics." Scu.edu, https://www.scu.edu/ethics/focus-areas/journalism-ethics/programs/the -trust-project/. Accessed 18 Feb. 2018.

27. Snow, Jackie. "Developers are using artificial intelligence to spot fake news." Business Insider, 17 Dec. 2017, http://www.businessinsider.com/spotting-fake-news-with-ai -2017-12. Accessed 18 Feb. 2018.

28. Edell, Aaron. "I trained fake news detection AI with >95% accuracy, and almost went crazy." Towards Data Science, 11 Jan. 2018, https://towardsdatascience.com/i-trained -fake-news-detection-ai-with-95-accuracy-and-almost-went-crazy-d10589aa57c. Accessed 18 Feb. 2018.

29. Newton, Casey. "YouTube will add information from Wikipedia to videos about conspiracies." The Verge, 13 Mar. 2018, https://www.theverge.com/2018/3/13/17117344 /youtube-information-cues-conspiracy-theories-susan-wojcicki-sxsw. Accessed 14 Mar. 2018.

Conclusion

1. "Democrats and Republicans differ on conspiracy theory beliefs." Publicpolicypolling.com, 9 Oct. 2017, https://www.publicpolicypolling.com/wp-content/uploads /2017/09/PPP_Release_National_ConspiracyTheories_040213.pdf. Accessed 7 Apr. 2018.

Index

Index